S0-BQM-361

PARANEOPLASIA

BIOLOGICAL SIGNALS IN THE DIAGNOSIS OF CANCER

PARANEOPLASIA
BIOLOGICAL SIGNALS IN THE DIAGNOSIS OF CANCER

Jan G. Waldenström, M.D., D.Sc.

Emeritus Professor of Medicine, University of Lund.
General Hospital, Malmö, Sweden.
Consulting Physician, Department of Oncology, Karolinska sjukhuset,
Stockholm, Sweden

A Wiley Medical Publication
John Wiley & Sons
New York • Chichester • Brisbane • Toronto

Library of Congress Cataloging in Publication Data

Waldenström, Jan Gösta, 1906–
Paraneoplasia.

(A Wiley medical publication)
Bibliography: p.
Includes index.
1. Paraneoplastic syndromes. I. Title.
[DNLM: 1. Neoplasms—Diagnosis. 2. Neoplasms—
Metabolism. QZ241 W162p]

RC259.W35 6,6.9'94'072 78-18494
ISBN 0-471-03490-8

Printed in the United States of America

10 9 8 7 6 5 4 3 2 1

To Karin

Foreword

Evidence has been accumulating for several years indicating that the growth of neoplastic cells is accompanied by the concomitant synthesis of many organic materials, particularly polypeptides. While most amino acids, peptides, glycoproteins and macromolecules, derived from abnormally proliferating cells, are probably degraded, salvaged, or excreted by normal mechanisms, some have well-defined hormonal or pharmacological activities and may be produced in abnormal quantities. When the concentration of these physiologically active compounds is great enough, clinically distinctive paraneoplastic syndromes develop. These clinical syndromes, which occur in patients with various forms of cancer, are, by definition, abnormalities that are not produced directly by the invasion of tissues by neoplastic elements or by mechanical disruption of normal organ function. Paraneoplastic phenomena refer to a variety of indirect and usually remote effects produced by tumor cell metabolites or other products.

Among the first recognized paraneoplastic abnormalities was the appearance of anomalous immunoglobulins, or globulin chains, in the plasma and/or urine of patients with lympho-plasma cell malignancies. Investigation of these proteins in man, and in experimental animals, has yielded important information regarding their genetic control, chemical structure, and pathways of normal immunoglobulin synthesis. An abundance of evidence indicates that perversion of normal cell function accompanies neoplastic cell transformation.

Many small molecules with pharmacological activities or other properties such as serotonin, glucagon, or lysozyme (muramidase), and unusual patterns of enzyme activity have now been identified in patients with neoplastic disease. Some of these abnormalities may represent tumor cell markers, and on occasion facilitate the early diagnosis of malignancy, while others may be important in the assessment of premalignant states, the quantitation of therapy, or the evaluation of tumor progression. The recognition of paraneoplastic phenomena has thus become relevant to the diagnosis and investigation of many everyday problems in clinical medicine and in oncology.

Professor Waldenström has written a timely, comprehensive, and easy-to-read monograph which summarizes a large amount of present-day knowledge regarding paraneoplasia, and identifies many areas in urgent need of further investigation. His extensive clinical investigative experience with a wide variety of diseases over the course of many years makes him uniquely qualified for this endeavor. From an historical standpoint, he points out that one of the early observations, which stimulated subsequent work in the field, was that hypercalcemia was sometimes associated with renal or ovarian carcinomas without bone metastases being demonstrable. This was soon followed by the demonstration that ectopic ACTH production occurred in some patients with carcinoma of the lung and produced hyperadrenalism. He uses the term "ectopic" to refer to the production of hormones, or other substances elaborated by tissues that are not normally concerned with their synthesis, in contrast to the "topic" production of monoclonal immunoglobulin by plasma cells in myeloma. The potentiality of different tumors to produce paraneoplastic substances is known to vary with their tissue of origin, function, ontogeny, and degree of tissue specialization. The production of active polypeptides by tumor tissues can be interpreted as the result of depression of dormant genomes.

The signs and symptoms of paraneoplasia may involve the skin, nervous system, bone marrow, lymphoreticular tissues, vascular system, kidney, skeleton or soft tissues. The specific organs and tissues involved are reviewed in several chapters which will be particularly valuable to clinical specialists and investigators. Some of the more common paraneoplastic syndromes describes are easily and widely recognized, but many rare entities may not be: acronecrosis, atypical amyloidosis, familial hyperkeratosis associated with carcinoma of the esophagus, hypereosinophilia with eosinophilopoietin, scleromyxedema and ostersclerosis with polyneuopathy in patients with plasma cell disorders, lactic acidosis in patients with acute leukemia and Burkitt's lymphoma, villous tumor of the rectosigmoid producing watery diarrhea and hypokalemia, sclerosing hemangiomata, glucagon and somatostatin secreting tumors, and many others.

The number of specific tumor-produced biochemical disorders is increasing rapidly and the most interesting, perhaps, are those associated with the production of topic or ectopic polypeptides. The synthesis and occurrence of these substances has some parallel with the formation of immunoglobulins by plasma neoplasia. A number of clinical syndromes may develop as a consequence of the monoclonal proliferation of immunoglobulins. Many of the paraneoplastic syndromes may be explained by the activation of one polypeptide-forming template.

This is a stimulating and informative monograph that deserves the careful attention of all clinicians and oncologists. The bibliography is extensive and provides many references to work that may not be familiar to American physicians. This monograph represents a scholarly landmark in an exciting field of medicine in which we expect rapid progress to occur in the years to come.

Wayne Rundles, M.D.
Professor of Medicine
Duke University School of Medicine
Durham, North Carolina

Preface

The problem of recognition and early diagnosis of malignant tumors has become increasingly important with the advent of active therapy. Ideally, the problem would be solved if some serologic reaction indicating the presence of cancer could be discovered. It is probable that a panel of different tests may become useful, but for now, it is necessary to learn as much as possible about the clinical signals from carcinomas. During recent years a large number of these signals have been recognized, and it is quite clear that the physician, who understands their meaning, has a chance to make the diagnosis of carcinoma at an early date. Some of these signs and symptoms are rather nonspecific. Others have been much discussed as true cancer signals or as possible connections. Therefore, I have tried to collect convincing proof—from my own experiences and from the literature—that in each case there is a real paraneoplastic mechanism that reverses after active treatment of the neoplasm. The present trend in medicine is to analyze a large number of cases and to collect many facts that may be treated statistically. On the other hand, it is clear that clinical medicine must have treatment of the individual patient as its chief goal. Therefore also the anecdotal observation may be of great importance, if judged critically.

The second problem has to do with the biology of cancer and the metabolic disturbances connected with the presence of tumors in the body. Analysis of paraneoplastic phenomena has demonstrated that a large number of completely different carcinoma cells have the faculty to produce active substances that cause recognizable clinical pictures. When the mechanism behind these syndromes has been analyzed, it often has been found that the cause of the disturbance is the formation by the cancer cells of different polypeptides which enter the bloodstream.

The true cause of this phenomenon is still unknown but the assumption is that many conditions will be explained when we look for specific polypeptides from the cancer. This may apply to such common symptoms as cachexia, anemia, and fever, as well as to the more special dermatologic or neurologic disturbances. The optimist might believe that the discovery of such toxic substances could lead to symptomatic treatment of the patient with improvement of his general condition even if the tumor itself is intractable. I believe that this theoretical problem—the cancer cell as a random producer of a number of active and also inactive polypeptides—is intimately connected with the fundamental question of the unlimited growth of the tumor cell.

Jan G. Waldenström

Acknowledgments

This book contains experiences from a lifetime in clinical medicine and in biochemical research. It has been an attempt to combine bedside and benchside. This means that I should like to express my gratitude and indebtedness to so many colleagues and friends with different interests and in many countries. It is impossible to mention them all.

Early experiences in Cambridge, England and in Munich, later collaboration with many research workers in Uppsala, Malmö and Stockholm as well as many friendships in the U.S. and other countries have given me a strong feeling of gratitude towards the scientific community as a source of inspiration and information.

If any names should be mentioned among my friends who are still with us I think that Kai O. Pedersen in the Institute for Physical Chemistry in Uppsala and Carl Bertil Laurell, Head of the Department of Clinical Chemistry in Malmö have a very important place in my development. Special thanks go to the past and present members of the staff at the General Hospital in Malmö for inspiration and fruitful collaboration in clinical medicine.

Oncology in Sweden is moving in new directions. During the last years I have had the pleasure to work together with Jerzy Einhorn and members of the staff at the Karolinska Hospital in Stockholm. This has been of special value for the writing of this book that I hope will become an introduction to clinical oncology. For the preparation of this book I have had excellent secretarial help from Ms. Birgit Lundgren, Birgit Hultman and Ann-Britt Nordui.

Contents

PARANEOPLASIA

BIOLOGICAL SIGNALS IN THE DIAGNOSIS OF CANCER

Introduction

Paraneoplasia (p.n.), a comparatively new term, literally means something that occurs beside the tumor and is somewhat ambiguous. Another term that is sometimes used synonymously, but erroneously, is ectopic hormone production. It is clear from recent experiences that paraneoplasia should be more inclusive even though the appearance of substances with well-defined hormonal activities is one, and perhaps the most striking, example of paraneoplasia. The crucial experiment to prove a causal relationship with the tumor would be the demonstration that extirpation of the tumor cures the paraneoplastic symptoms. Therefore, I have tried to collect as many instances of *reversible* paraneoplasia as possible from the literature and from my own experience. In doing this, it has become obvious that well-described and convincing cures with sufficient follow-up are quite rare in the literature. This is due partly to the fact that clear-cut paraneoplastic symptoms often develop at the stage of the neoplastic disease when numerous metastases are already present, and partly to the fact that many authors publish their cases much too quickly without awaiting the final development.

HISTORY

The history of p.n. is quite interesting, and several authors have been cited as having realized the causal connection between some seemingly odd symptoms and the presence of a malignant tumor in the same patient. Fuller Albright was probably the first person who saw the possibility that a tumor may form a special, metabolically very active substance even if the source of the tumor does not indicate that it may have any metabolic activity. Albright observed a patient who had a renal carcinoma and hypercalcemia and showed no clinical signs of osseous metastases. He hypothesized that the tumor may have produced a calcium mobilizing factor (3). On careful examination, one skeletal metastasis was found. At that time it was regarded as self-evident that patients with bone metastases and hypercalcemia had the latter symptom as a result of "mechanical" bone destruction and subsequent calcium mobilization.

It is interesting to note how the pendulum now has swung in the opposite direction. Recently it has been demonstrated that even patients with widespread bone metastases and hypercalcemia may have high levels of parathormonelike substances in their blood. Plimpton and Gellhorn at Columbia University were the first to show that hypercalcemia in tumor patients without bone metastases may be reversible after extirpation of the tumor (339). They observed a patient who had ovarian carcinoma and very high serum calcium levels. Surgical exploration of the neck did not disclose any parathyroid adenoma, and the patient's condition remained unchanged until the extirpation of the ovarian carcinoma. This led to rapid but only transient normalization of the serum calcium level. After some months, the calcium values increased and at the same time metastases developed.

ECTOPIC AND TOPIC PRODUCTION

Liddle and his colleagues coined the expression ectopic hormone production in 1962 (282). They were able to demonstrate that some patients with classical Cushing's disease had a carcinoma that was not in any way directly connected with the adrenal glands or the pituitary, and that the endocrine disease improved after extirpation of the carcinoma. It was thus clear that the cancer cells produced some substance that caused the clinical picture of Cushing's syndrome. These patients also had bilateral hypertrophy of the adrenal glands, and Liddle hypothesized that this was caused by increased adrenocorticotropic hormone (ACTH) production by the tumor. Clearly the production of ACTH by cancer cells from an organ such as the lung must be regarded as "ectopic," since this hormone normally is produced by cells in the pituitary gland. Hence, the now generally accepted term ectopic hormone production is applied to many tumors in different organs.

It seems important to make a distinction between such ectopic formation of active substances (hormones) and the "topic" production in tumor cells arising from an endocrine gland, whose function is to synthesize this active molecule, even if those cells are dislocated or "ectopic" anatomically. The famous Viennese surgeon von Eiselsberg published the history of a patient with thyrotoxicosis. After complete thyroidectomy, the symptoms disappeared, but it was clear that the goiter contained carcinomatous tissue. After some time the patient had a relapse and simultaneously developed a metastasis in the sternum. Since that time numerous similar examples have been collected. They prove that the tumor cells, e.g., in the hepatic metastases of endocrine tumors, may have the synthetic apparatus needed to produce active hormones "topically" on the cellular level and ectopically on the organ level.

I shall try to keep the distinction clear between topic and ectopic p.n. and chiefly discuss the latter. Topic production of active substances will be treated in this work only when active substances are produced that may be recognized through their clinical activity and therefore lead to correct diagnosis of the tumor.

The most important symptoms of malignant tumors such as cachexia, anemia, and fever in all probability are caused by "toxic substances." Modern oncology will have to attack these problems with biochemical methods. When these "toxins" have been isolated and defined chemically, the ways to neutralize their action probably will increase tremendously. There can be no question that certain tumors, even if widespread and consisting of enormous quantities of tumor cells may cause comparatively few general symptoms, whereas other smaller tumors may cause death rapidly. It is regretable that as yet we cannot fruitfully discuss the chemical basis for such symptoms. Discussion of p.n. may awaken our interest in these phenomena.

ECTOPIC HORMONE PRODUCTION IN MALIGNANT TUMORS

The reader may well ask why *hormone* production should be of special interest since it may be assumed that tumor cells produce a number of topic or ectopic substances that are not recognized as hormones. This is also true. However, this special interest is explained by the fact that some hyperhormonal conditions have well-known clinical pictures. It is easy to make the diagnosis of ACTH production by a malignant tumor when the patient has cancer and Cushing's disease, especially if he is also hyperpigmented. Several syndromes with increased gonadotropin production are also comparatively easy to recognize. We have reason to believe that malignant tumor cells produce large numbers of different substances that are ectopic. Their presence can only be suspected, and the suspicion must be confirmed chemically if the substance has no known clinical activity. Fetuin in liver carcinoma is a classical example. It is produced by otherwise "resting" (derepressed) embryonal systems.

In 1964 I pointed out that tumors produce active paraneoplastic substances that are always and only polypeptides, and I discussed the fact that such activation of protein-forming templates might be characteristic of malignant cells (464). This point of view was further elaborated in the Flexner lectures that were held in Nashville in 1965 and published in 1968 (456). I compared ectopic ACTH production and the production of monoclonal immunoglobulin by myeloma cells. The first is ectopic; the second, topic; but both are probably the result of derepression of specific protein-forming templates (see also 465).

It is clear that many of these active paraneoplastic substances are produced as truly ectopic products, but it is also evident that some tumors have a greater potentiality for producing a variety of active polypeptides than do others. This fact is easy to accept if one favors the theory of derepression. It seems very plausible that tumors originating from a pluripotent stem cell, such as the neural crest, should have a greater propensity to synthesize a number of active hormones than do neoplastic cells from a more highly specialized tissue.

Masson was interested in the origin of the argentaffine cells in the intestinal wall and regarded them as part of the nervous system. Later the German pathologist Feyerter wrote extensively on the "helle Zellenorgan." He found such light cells in a large number of organs, but his idea was to unite them into an organ. He was not, however, at that time conscious of the products (138). In Britain Pearse and Pollack have worked out the concept of the APUD cells in great detail, and they point out that these are derivatives of the neural crest and have retained the faculty to form both polypeptide and amine hormones (329). In some instances this double synthetic capacity is very striking: The carcinoid tumors arising from argentaffine cells produce 5-hydroxytryptamine (5-HT) by decarboxylating 5-hydroxytryptophan, and it is believed that they also produce kinins, i.e., polypeptides that are highly vasoactive. In contrast, medullary thyroid carcinoma produces a polypeptide, calcitonin, but is hardly a regular producer of any amine (see Chapter 10), and pheochromocytomas produce catecholamines, but as far as we know, they do not normally produce active polypeptides.

Recent work has demonstrated that the bronchial carcinoid cell is probably closely related to cells in thymomas and in oat-cell carcinoma. The fact that the carcinoid cell is one of the most pluripotential tumor cells may explain why oat-cell carcinoma is a common cause of complicated paraneoplastic syndromes. Another pluripotential cell is the pancreatic islet cell. Even if the beta cells are usually connected with insulin production and the alpha cells with glucagon, it is quite clear that malignant tumors from islet cells have a wide capacity for production of polypeptide hormones. Very recently, the so-called D-cells have been found to contain large amounts of somatostatin, a new polypeptide hormone with interesting functions (191). The number of active polypeptides present in the intestinal wall -or in the islet cells of the pancreas is very large, and their biochemical connections are still under investigation (80).

Even if it is clear that tumors consisting of certain cell types, such as those derived from the neural crest are especially prone to produce ectopic polypeptides, it cannot be denied that there are instances of ectopic production in tumor cells that have nothing to do with the normal synthesis of the ectopic product. Several attempts to explain these situations have been made. The assumption is tempting that cell fusion may occur between the topic and ectopic cells in such a way that synthetic templates from the topic cell are incorporated into the ectopic cell with malignancy developing later. (Weichert [485] gives a very interesting discussion of fundamental problems.) The omnipotentiality of the genome in every cell—*even the somatic*—is demonstrated in Gurdon's experiments. He has been able to show that the naked nucleus from a somatic (intestinal) frog cell may be incorporated into the ooplasm of an enucleated egg cell, and in this environment the nucleus starts to divide meiotically. In a few nuclear transfer experiments Gurdon succeeded in producing a complete fertile frog! This seems to be a completely convincing proof that the genome in a somatic cell is only dormant and may be awakened by the right stimuli. Therefore, it is a strong argument in favor of the assumption that derepression may occur also in carcinoma cells.

It may be worthwhile to draw some parallels between the action of topic hormones such as ACTH and thyrotropin (TSH) and the influence of an antigen on a clone of cells carrying the corresponding antibody molecule. Both contacts lead to synthesis of a special kind of protein and, at the same time, to cell growth. It is now clear that there are dormant genes for the different parts of the immunoglobulin molecule in the immunocyte. Each cell clone only derepresses the templates consisting of several genes coding for one whole immunoglobulin molecule. It is now known, that even the light chain with a molecular weight of 25–30,000 daltons is coded by at least two genes. Many facts favor the assumption that p.n. caused by peptide formation only (or usually?) happens with respect to molecules that are produced on one ribosome. Thus, the activation may occur at this point, and it seems probable that the growth of the cancer cell and activation of polypeptide forming ribosomes are in some way connected.

We have tried systematically to find patients who produce light chains of immunoglobulin molecules as a paraneoplastic phenomenon. So far neither we nor anybody else have been able to find a carcinoma connected with Bence Jones proteinuria and still less with the formation of the whole antibody molecule. Even in such cases where no definite myeloma is present, we have strong indications that the immunoglobulin production of monoclonal nature occurs topically in plasma cells spread in the whole bone marrow (Hijmans, Turesson). This is a benign plasmocytoma or a benign monoclonal gammapathy.

Appendix 2 contains data on possible paraneoplastic products.

CELL CULTURE

One of the best indications that a tumor is producing the hormone in question is the incorporation of labeled amino acids into the hormone molecule produced by tumor cells in a culture. The first experiment of this kind was probably performed by Klein et al (237). These authors obtained tumor tissue from a 45-year-old man who suffered from an inappropriate secretion of antidiuretic hormones (SIADH) and had an oat-cell carcinoma of the lung. Both labeled tyrosine (^{14}C) and leucine (^{3}H) were added to the medium. Vasopressin contains tyrosine but not leucine, while oxytocin contains both. Only carbon 14-labeled vasopressin was demonstrated in the cultured cells, along with a leucine-containing material that reacted with the antiserum.

Two other antidiuretic hormone(ADH)-producing tumors have been examined, both with labeled phenylalanine (^{3}H).

George et al studied a 68-year-old man with marked leukocytosis but no anemia (152). His serum sodium was 119mEq/L; chlorides, 84 mEq/L; potassium, 4.3 mEq/L; and creatinine, 0.9 mEq/L. Urinary sodium was: 110 mEq/L in 24 hours, and the specific gravity was 1.014. The patient had a pulmonary tumor that was extirpated. His hyponatremia was treated by saline infusion, but the condition was hardly normalized. The tumor was diagnosed as bronchogenic carcinoma, the cells of which contained electrondense granules and large mitochondria. Cell culture demonstrated the incorporation of labeled phenylalanine into a substance that was identified as arginin vasopressin on column chromatography.

Similar results have been reported by Martin et al. In this case the tumor was an anaplastic carcinoma from the lung.

Greenberg has published an excellent study on two carcinomas with paraneoplastic syndromes (169). One was an undifferentiated lung carcinoma that incorporated ^{14}C leucine into a substance that was identified as HGH. The second was a clear-cell renal carcinoma. In the latter case the patient had an elevated serum calcium (11.2–12.0mg/100 ml), low phosphorus (2.8–3.3 mg/100 ml), slightly elevated creatinine level, an elevated alkaline phosphatase level, and an elevated plasma PTH level. Within 12 hours after nephrectomy the serum calcium level returned to normal, and the plasma PTH became undetectable. Extracts of the tumor were investigated by radioimmunoassay. No PTH was detected. Cells from the tumor were cultivated in a medium containing labeled leucine (^{14}C) and selenomethionine (^{75}Se). The authors had previously worked with cultured human parathyroid cells and found that this immunoreactive PTH was indistinguishable from bovine PTH. In the tumor cell culture, the immunoreactivity differed from bovine PTH with two of three antisera. Riggs had found a similar difference between ectopic and topic PTH. It is possible that fragments of the molecule are released, or that pro-PTH may be released by the tumor cell. The authors also discuss the possibility of immunochemical heterogeneity. Other discrepancies between biological and immunological activity have also been noted by previous authors.

GRADIENT

Another important argument in favor of genesis in the tumor is the establishment of a hormone gradient through the tumor. During surgery, it may be possible to compare ingoing arterial blood and outgoing venous blood. If the hormonal levels are higher in the latter, this has been regarded as a strong proof of hormone production in the tumor. Extraction of the specific substance from specimens of tumor is also important. In all experiments on hormone gradients the specificity of the methods used is of course decisive.

Buckle et al observed a 58-year-old woman with a renal tumor and hypercalcemia (serum calcium, 12–13.2 mg/100 ml). Oral cortisone did not lower the serum calcium level. The tumor was an adenocarcinoma with clear and eosinophilic cells. The patient was febrile and had an increased ESR, but the fever disappeared and the ESR returned to normal after surgery. Concentrations of PTH in the right renal artery was 0.6 and in the right renal vein, 2.5, which seems to be a convincing gradient (70). Williams et al showed a similar gradient through the tumor in their patient with liver carcinoma that produced PTH . The PTH values were higher in hepatic venous blood than in portal or hepatic arterial blood (see also 239).

BIOCHEMISTRY

One of the many reasons why it is difficult to decide if a polypeptide from a tumor with hormonal activity is the same molecule as the normal hormone is that there is great polymorphism among polypeptide hormone molecules. In 1968 Yalow and Berson investigated the normal PTH molecule present in plasma and in extracts from the glands. Such studies demonstrated that the parathyroid hormone also occurs as a much bigger molecule than was previously accepted (507). This finding has started a number of investigations into the molecular size and polymorphism of several protein and polypeptide hormones. It is probable that such a heterogeneity is present in a large number of polypeptide hormones, and it has been proven for several. There are many possible explanations of this phenomenon. Early studies made on ACTH showed that a large part of the molecule could be very active and perhaps still not detectable by immunoassay. It is therefore important to test not only the immunological but also the physiological reactivity. This is difficult, however, because tumors do not produce large amounts of these substances. Sequence analysis of amino acids requires very pure preparations.

The subunit story has become very important in discussing paraneoplasia. It is well known by now that the gonadotropin molecule can be split into two nonidentical chains that have been called alpha and beta. It appears that the alpha chain is the same in all hormones that are glycoproteins, whereas the beta chain is specific for the activity. Alpha and beta chains show no cross-reactivity. Recombinations produce an increase in activity. Of particular interest is the formation of only alpha chains in patients with cancer. Several authors have succeeded in demonstrating the presence of such units by immunoassay. This shows that hormones without activity have been formed by tumor cells. Instances are known with a high content of alpha chains being present with or without complete FSH, LH, or TSH (see Chapter 11). It is evident that the two chains may be synthesized independently. The parallel with partial synthesis of immunoglobulin molecules in myeloma patients is obvious. Most myeloma cells synthesize the whole molecule. Some *also* excrete the light chain, whereas a number of cases only have an excess formation of light chains.

If we assume that only topically secreted hormones may be formed on several templates, whereas ectopically produced polypeptides should be the product of one ribosome, we may ask if the two chains in the gonadotropin hormone molecule may ever be formed ectopically. The same is true of immunoglobulin molecules that do not seem to be formed by cancer cells. For several years it was difficult to understand why the seemingly simple light chain (Bence Jones protein) was never formed by cancer patients in the absence of plasmocytoma. Recent work on the synthesis of light chains has shown that the constant part and the variable part on the molecule are obviously synthesized on RNA coded by different genes. The question of insulin synthesis has always been the subject of heated arguments. Steiner has provided a completely new basis for this discussion. It was believed that the A and B chains of the molecule were probably synthesized separately. It has now been proven beyond a doubt that insulin is formed as a big proinsulin molecule that is split and rearranged to form the insulin molecule. We assume that proinsulin is formed on one ribosome. The whole problem is further complicated by the presence of substances with insulinlike activity (ILA). It is probable that thorough investigation of sera from patients with severe hypoglycemia without insuloma will give the ultimate answer to these problems. The nature of ILA is still unknown and may be important for this discussion.

Normal PTH has 84 amino acids. It is probable that pro-PTH has 109 amino acids with a molecular weight of around 12,000 daltons. It has not yet been established if pro-PTH is the normal product of the gland. Some authors believe that the concept of microheterogeneity should also be applied to polypeptide hormones. We know that several preparations of one enzyme contain active units with different electrophoretic mobilities. These are called isoenzymes. It is possible that isohormones may be formed and Keutman studied this problem in detail when he isolated three different PTH molecules (232).

The production of fetuin seems to be an important argument in the discussion of the mechanisms for the synthesis of paraneoplastic polypeptide chains. This synthesis is obviously being repressed during the early phase of extrauterine life. The templates for synthesis are located in the liver during fetal life, and it is therefore natural that tumors from liver cells should be the chief source, when tumor derepression occurs. The production is topic but may be called asynchronous.

Another polypeptide synthesized during fetal life and later repressed is the fetal chain in the hemoglobin F molecule. One patient with carcinoma and a high content of hemoglobin F has been observed in Malmö. If there is a causal connection, the most probable mechanism would be formation of a factor that "turns on" the synthesis of the chain. It might be rewarding to look for hemoglobin F in malignant erythremia (di Guglielmo's disease). Regarding fetal hair in carcinoma patients see Chapter 1, Section 14.

Part 1

SIGNALS FROM ORGAN SYSTEMS INDICATING THE PRESENCE OF NEOPLASIA

1
Skin

1. Flushing
2. Erythema Gyratum Repens
3. Erythema Annulare Centrifugum
4. Necrolytic Erythema
5. Pyoderma Ulcerosum Serpiginosum
6. Paraneoplastic Acrokeratosis (Bazex)
7. Hyperkeratosis
8. Acronecrosis (Hawley)
9. Ichthyosis
10. Pemphigoid
11. Porphyrin Production by Tumors
12. Acanthosis Nigricans
13. Pigmentation and Cancer
14. Hypertrichosis
15. Metabolic Disturbances (connected with Melanoma)
16. Dermatological Symptoms Connected with Myeloma, Macroglobulinemia, and Amyloidosis
17. Dermatomyositis
18. Bowen's Disease

For obvious reasons the skin should be an excellent mirror of processes occurring within the body. A number of skin lesions are more or less characteristic of some important malignant processes, on the other hand some of these dermatological symptoms indicate very rare types of neoplasia, e.g., histamine type of flushing in gastric carcinoid or precocious puberty in boys with pinealoma. Therefore they do not usually have very great practical importance, even if they may open up new vistas for our understanding of biological processes. Certain changes in the integument are rare even when they occur in tumors that are in themselves not uncommon. Examples of this are acanthosis nigricans in gastric carcinoma or generalized lanugo appearing acutely.

1. FLUSHING

Among the many external paraneoplastic symptoms "phenomenal flushing" is one of the most striking. It occurs in a number of conditions, but the most important is the malignant carcinoid tumor (see Chapter 10, Section 1). In this disease several types of cutaneous hyperemia may be observed. The least striking is a simple slightly cyanotic hue, which may be regarded as constant until it disappears momentarily after complete extirpation of the tumor. Much more dramatic is the paroxysmal flushing that usually lasts for some minutes and often may be localized not only in the flush area (face and upper thorax) but also on the trunk and extremities (474). The most common finding is a bright red coloration stretching over large surfaces and lasting for minutes. It is not really patchy but tends to have indistinct borders. In a few patients I have seen rapid changes between erythema, cyanosis, and a more salmon-colored skin (for colored pictures, see 460 and 472). These grand flushes are accompanied by borborygmus and usually a strong urge to defecate. They may be provoked by different stimuli and are most common and severe after the morning meal. Alcohol and strong-tasting cheese also may induce such attacks. It is quite clear, however, that purely psychological factors are also active. Most patients tend to become flushed during rounds or when they are demonstrated at medical meetings. This flushing is probably mediated via nervous stimulation of the adrenal medulla since catecholamines regularly release the substance responsible for flushing. It usually is maintained that paraneoplastic release of active substances is not mediated by nervous impulses since there is no innervation in tumor tissue. The carcinoid flush is no exception since there is certainly a humoral factor (catecholamines) acting directly on the argentaffine cells. There is a general redistribution of blood during the flush that we also have seen on the peritoneum during a laparotomy (for a further discussion, see 437 and 438).

A second type of flushing is seen in carcinoid patients with a tumor located in the foregut. We observed a patient with a tumor producing osteosclerotic bone and lung metastases who had a type of flush that resembled the histamine flush. The flushing lasted much longer—up to an hour (476). It was patchy with irregular borders, very deep red, and itching. At the same time there was intense hyperemia of the conjunctiva and the eyelids were swollen (for colored photographs, see 472). This patient produced large amounts of histamine, and it is clear that this is a real histamine flush. There was no hyperperistalsis. Later it was found that this patient had a gastric tumor. (For a discussion of the biochemistry in carcinoid tumors see Chapter 10, Section 2).

The most common type of flushing is seen in patients with climacteric complaints. These women may perceive their flushes as being intensely hot, even when their skin is not very red. I have had the opportunity to watch some women closely when they have had their flushes. It is usually possible to tell the patients when they are having a flush, but sometimes they feel their "hot spells" without such visible erythema. Most carcinoid patients feel when a flush is coming. Menopausal flushes should not cause any diagnostic difficulties.

In the last few years there have been published reports of flushing occurring in patients with three other types of tumors. One type is medullary thyroid carcinoma, where diarrhea is also common, and in my experience with such patients, flushing is seen chiefly in severe stages when diarrhea is predominant (see Chapter 10). There is no time connection between these two conditions, however. Flushing is seen in patients with this tumor who do not have an increase in 5-HT. The second type—neuroblastoma—is much rarer, and occurs in

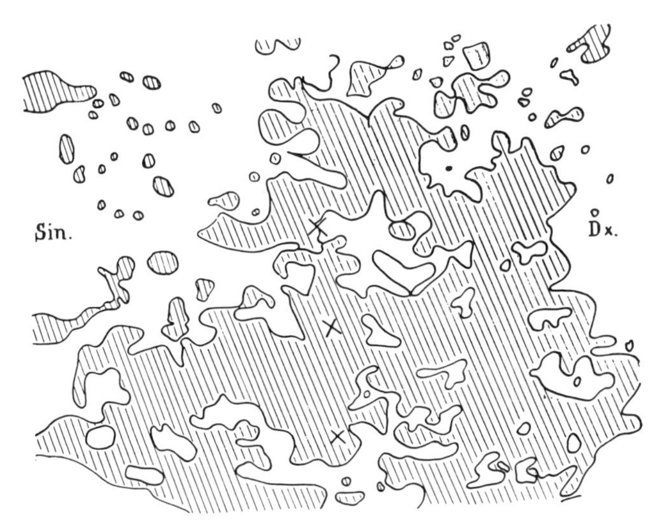

Figure 1. Skin on abdomen during maximal flush. The white skin is of normal color; the hyperemic parts, bright red.

children. (see Chapter 10, Section 4). The third type—a special form of an alpha-cell pancreatic tumor, that often metastasizes to the liver—is very rare and may cause severe diarrhea and paroxysmal flushing. This is called pancreatic cholera. There is good evidence supporting the causal role of vasoactive, intestinal polypeptide (VIP) in this condition (for further discussion see Chapter 11).

A. Acrocyanosis

This symptom is not a real flushing even if the skin colors vary much from time to time. This subject is treated in Chapter 9.

2. ERYTHEMA GYRATUM REPENS

One of the most fascinating paraneoplasias of the skin is without doubt a condition that was first described by Gammel under the name erythema gyratum repens (148). Interestingly enough this condition is only seen in the presence of a carcinoma, and the successful treatment of the tumor results in complete disappearance of the skin lesions. In several cases the first manifestation of the dermatologic symptom has been rather atypical with erythema, scaling, and itching of the face. Later a rash develops on the trunk and proximal limbs. The periphery of each lesion often shows scaling, and the borders are serpiginous with normal skin between. The pattern changes from day to day, and the advancing front of the gyri has been measured in different patients and has been seen to advance about 1 cm per day. The pattern on the skin may be very configurate, and some authors have described a zebralike appearance. It is common with severe itching and some patients have exhibited a high degree of eosinophilia.

Purdy saw a 45-year-old woman with diffuse ductal carcinoma whose first symptom was severe itching over her right breast. Within 24 hours after mastectomy the itching disappeared. The erythema disappeared completely by the third day, but severe desquamation lasted for a week. An interesting observation was made by Pevny (337).

A 64-year-old woman developed typical dermatologic symptoms in May 1964. In August of the same year she had a profuse hematemesis, and gastric cancer with liver metastases was found. Previous GI roentgenograms, however, had been normal. The patient died in November and it was noted that during the last weeks before death, the dermatologic symptoms were not apparent.

Pevny discusses in great detail the differential diagnosis between erythema annulare centrifugum, a rare sign of malignancy, and erythema gyratum repens, which is always a herald of neoplasia (337).

At this time, nine carcinoma patients with different carcinomas were known: two mammary, one bronchial, one hypopharynx, one collum uteri, one cervix uteri, one tongue, one cardiac, and one gastric carcinoma. In all the skin symptoms had been very similar.

In 1970, Thomson and Stankler described two patients, both of whom were initially diagnosed as having pityriasis rubra pilaris but whose lesions later developed a typical changing gyrate pattern. One of the patients had a cancer of the vesica urinaria, the other was examined very carefully for abdominal neoplasia and underwent an exploratory laparotomy. The results were negative. Six years after the skin symptoms developed, a prostatic carcinoma was diagnosed (435).

Hochleitner et al observed a man with typical erythema gyratum repens over his whole body. He also had hyperkeratotic thickening of his palms and plantae, widespread enlarged lymph glands that were interpreted as lipomelanotic reticulosis, and pigmentations on the inside of the buccae. The patient died from an inoperable squamous cell carcinoma of the lung (200).

In 1964 Summerly, in his paper, "The Figurate Erythemas and Neoplasia," gave a long list of names for all these migrating erythemas, and he believes that distinctions are necessary (427). The three cases that are presented in the paper are not carcinomas.

The first patient had an untreated myeloma, and he had typical centrifugal erythema on his arms and on the right side of his neck. The advancing borders were bright red with cordlike elevations. The clearing central areas were scaling and pigmented. The patient had new lesions in other locations. The myeloma was not treated. The second patient had an acute leukemia and had oval, erythematous, sharply defined lesions on each forearm and on her left leg. The third patient had an ovarian cystadenoma and an erythematous patch with gyrate borders over the left deltoid. After operation the skin eruption disappeared.

I have already mentioned that Pevny discussed the difference between erythema annulare centrifugum and e.g.r. and pointed out the different importance of these two manifestations for the diagnosis of cancer (337).

3. ERYTHEMA ANNULARE CENTRIFUGUM

This is an uncommon skin disease that is probably related to erythema gyratum repens (See the color plate indicating different, paraneoplastic, reversible symptoms from the skin [reproduced with permission of the Acta Med. Scand and the authors 243a, 1978]).

In 1968 a high sedimentation rate was noted in a 56-year-old woman with myeloma. In 1969 she developed big erythematous patches on the upper part of her body and arms. These plaques were slowly growing, showed central healing, and had slightly elevated borders. Later she developed similar symptoms on her cheek and trunk. The largest lesions were on her trunk and were up to 7 cm in diameter. They healed spontaneously without leaving any changes in the skin, but new lesions developed continuously. The patient experienced slight itching. Because of the elevated ESR electrophoresis was performed, and an IgA-L of 2.8 g/100 ml was found. X-ray of the skeleton showed a normal picture. Sternal puncture showed a slight increase in plasma cells. As the diagnosis myeloma was regarded as somewhat uncertain, we only followed the patient's course for two years and saw that she continuously developed big erythema patches as before. Later the patient had some pain in the chest, and new x-ray films showed a large number of irregular confluent foci in the skull and small lesions in several ribs, which were not present 2 years earlier. She was treated with Alkeran and after 2 months the skin condition improved greatly. She has been completely free from skin symptoms for 3 years and her myeloma globulin has disappeared (see also 179 and 243a).

Patient G.A. (Figure *b*) on the color plate had a prostatic carcinoma. After treatment the erythema disappeared.

4. NECROLYTIC ERYTHEMA

Becker et al were the first to find that a patient with a widespread necrolytic rash had an islet-cell carcinoma. The patient had lost much weight and had a normochromic anemia, a sore red tongue, and diabetes. Twenty-five years later a second case with an identical picture was described by Church and Crane (83). These authors regarded the eruption as unique since it consisted of an annular erythema (combined with superficial necrosis of the skin), which was brief in duration and moved from one area to another. In this condition, there is a complex pattern of erythema, crusting, and normal skin, and sometimes superficial bullae may arise in pressure areas (see Chapter 8, Section 5).

In 1973 Warin described a patient who had been diagnosed as having eczema on her back that had spread to the trunk, arms, and legs. On examination she had a circinate, gyrate pattern of erythema with scaling and crusting at the edges and central clearing. Her tongue was red and smooth. Her serum iron level was low; TIBC, normal; hemoglobin, 100 g/L. This patient died, and at autopsy a large tumor in the tail of the pancreas with numerous metastases was found. Histologically it was an islet-cell tumor (482).

Harman mentioned a 70-year-old man who had had similar skin lesions for 12 years. This patient had intermittent glucosuria, anemia, and weight loss. His tongue was sore, smooth, and red. Laparotomy was performed, and the pancreas showed signs of chronic pancreatitis. No cancer was found in the lymph nodes. The patient died, and postmortem examination showed both a carcinoma in the tail of the pancreas and cystic pancreatitis.

Wilkinson described a 52-year-old woman who had been treated for several years for cycles of annular erythematous eruptions on her feet, legs, and central trunk. She also had a striking perioral dermatitis, X-ray films of the spine showed metastases. The skin lesions were described as being very similar to erythema gyratum repens, but the important difference is that necrolysis with partial epidermal separation was present (492).

5. PYODERMA ULCEROSUM SERPIGINOSUM

This condition has been described in ulcerative colitis for many years. In 1964 a group of German dermatologists observed the condition, also called dermatitis ulcerosa, in multiple myeloma (360). Six patients had chronic deep ulcerations on their lower legs. The sera from four of these patients contained monoclonal globulin fractions all of the IgA type. In three of the four patients no myeloma was found, but the fourth was considered to have the disease. In a very large sample of patients with myeloma, we have only seen one instance of pyoderma ulcerosum serpiginosum. (Colorplate)

The patient, a 75-year-old woman, developed a painful ulceration on her right lower leg, that did not heal. She was found to be severely anemic. Electrophoresis showed a small IgA kappa component. On x-ray examination a carcinoma of the colon was found. The patient was first treated with cortisone, but later the carcinoma was removed. The ulcers did not heal. Continued treatment with steroids seemed to improve the skin condition very slowly. Treatment with Alkeran was started, and she was given five blood transfusions. After several months her skin condition had almost completely disappeared and she had gained weight. Her monoclonal IgA disappeared completely, and the serum albumin level had improved. It is impossible to say if this patient had a myeloma or only a benign monoclonal gammapathy. Her condition has remained excellent (see 211, 217, 238, 245, 402).

6. PARANEOPLASTIC ACROKERATOSIS (BAZEX)

This condition is usually named the Bazex syndrome (29, 105). The lesions are seen only in men who usually have neoplasms in the bronchial (mainly in the oral cavity and the larynx, rarely in the lung) and GI tracts. Erythema, often of a violaceous color, and hyperkeratosis with adherent scales and pruritus is common. Rubbing may cause bleedings. The predilection sites are the back of hands, ears, nose, chin, feet, knees, and elbows. The nails often show changes at an early stage. Longitudinal striation is common but apparently very unspecific. Thickening of the whole nail is also seen and the nail may be partially destroyed. Total onycholysis is said to occur. It is found predominantly in patients between the ages of 38 and 72 years. In 50% of the cases, the cutaneous diagnosis preceded the finding of the cancer. Generally treatment of the neoplasia causes regression of the dermatological symptoms. A few rather detailed descriptions of this process have been published, but no case was followed until complete healing occurred (see also 112).

7. HYPERKERATOSIS

Howel-Evans et al studied two families with the interesting combination of tylosis palmaris et plantaris and a very high incidence of esophageal carcinoma (205). Of the 48 family members with tylosis, 18 developed this type of cancer; whereas among the 87 nontylotic members, only one had cancer of the esophagus, and it is possible that this patient had very slight tylosis. The combination is inherited as an autosomal dominant trait, and it is evident that the combination with cancer

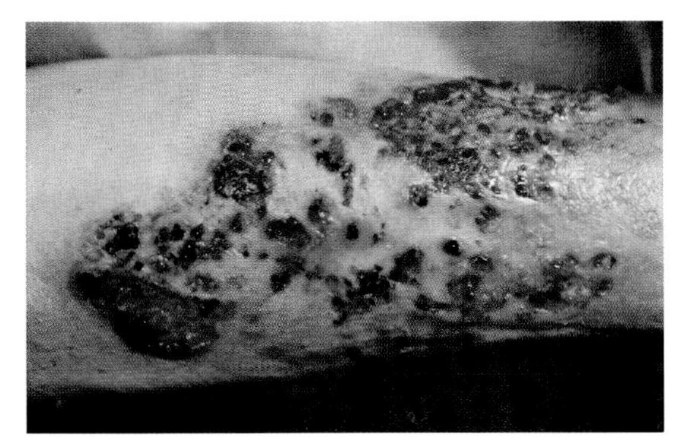

Fig. a. Pyoderma gangraenosum from patient E. L.

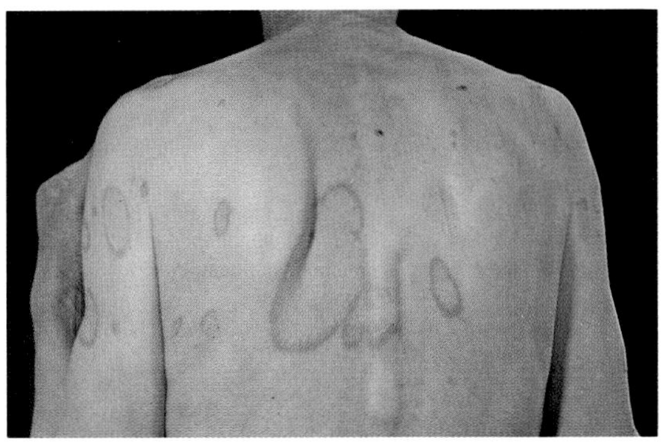

Fig. b. Erythema annulare centrifugum from patient G. A.

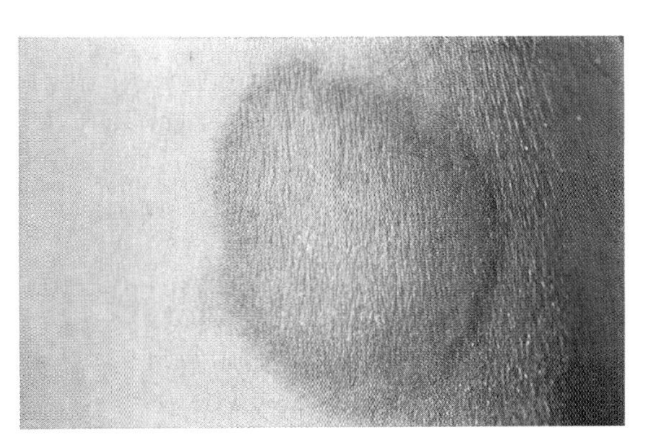

Fig. c and d. Erythema from patient E. H.

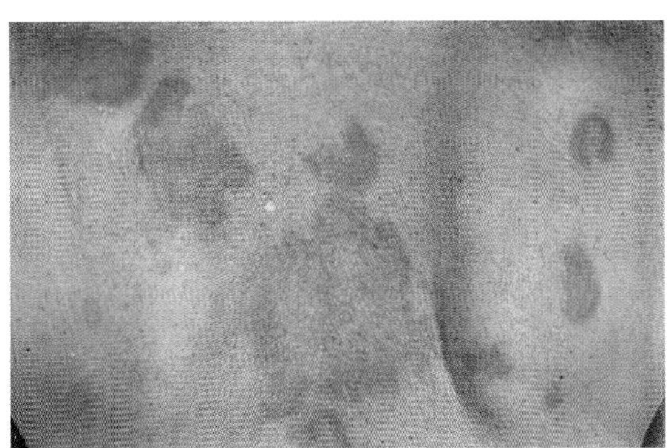

Fig. d.

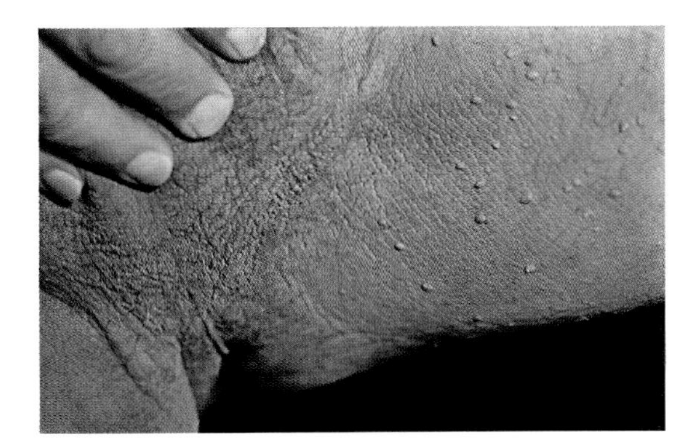

Fig. e and f. Acanthosis nigricans from patients described in the text.

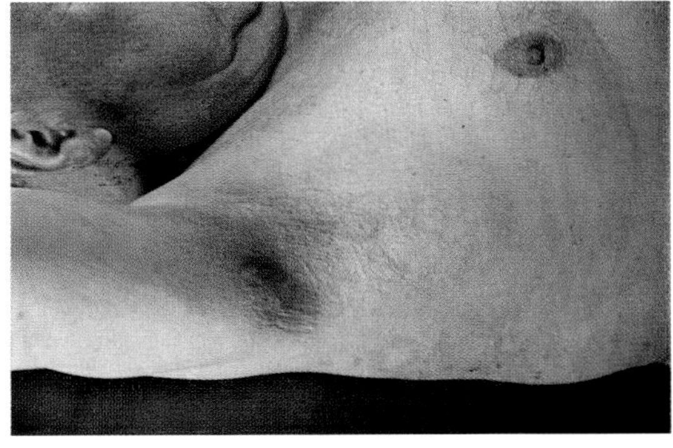

Fig. f.

occurs only in these two families; it has never been observed in a large number of members of other families with tylosis. Therefore, the type of tylosis found in these two families must be different from other types such as the so-called Meleda disease that is so common on that Greek island. Metabolic disturbances in epithelium of the hands, soles, and esophagus must predispose to esophageal cancer. In these families hyperkeratosis is a cutaneous signal of cancer, not a true paraneoplastic phenomenon.

8. ACRONECROSIS (HAWLEY)

In 1967 Hawley et al from University College Hospital in London published a report of six cases of Raynaud's syndrome with carcinoma (188). All patients were middle-aged women. The findings were bilateral in five patients and gangrene developed rapidly in four. In three patients the primary carcinoma was in kidney, ovary, or maxillary sinus. One patient had a metastases with no certain primary carcinoma (possibly ovarian or pancreatic). The last patient had Hodgkin's disease. All patients died rather rapidly, and in several the tumor was not detected before postmortem examination.

In 1976 Andrasch et al published observations on a patient with renal carcinoma who had suffered from severe Raynaud's syndrome (8). After radical nephrectomy for a sarcomatoid renal adenocarcinoma, there was rapid improvement in the circulation in her fingers. The ulcers healed in 2 months. This seems to prove that this type of acronecrosis is a true paraneoplastic syndrome.

We have seen one case that may have been an example of this curious, and certainly paraneoplastic, syndrome.

Man, born in 1909. The patient had been an alcoholic for many years. In June 1968, he complained of cramps in his legs and pains in his fingers. His fingers became severely discolored and bullae and ulcers developed. The radial pulsations were normal, but on arteriography changes in the arteries of the small fingers were found and interpreted as indicating arteritis. Somewhat later red papules were seen on both lower arms. The tentative diagnosis of Raynaud's syndrome caused by cryoglobulins had to be abandoned since very careful examination did not show any cold precipitable protein.* On admission the patient had very severe pain in the fingers. With symptomatic therapy his pains improved somewhat, and many of the ulcers healed. It was evident that the patient had a typical liver cirrhosis with water retention, ascites, hypoalbuminemia, and esophageal varices. He was considerably improved and, in December 1968, was released on a regimen of prednisone. After 2 months he was readmitted with severe liver failure and showed clinical signs of a malignant hepatoma. He died after being in an hepatic coma for 5 days. The postmortem finding was cirrhosis with carcinoma.

Cryoglobulinemia and cold agglutinin disease may cause disturbances in

*To ascertain the presence or absence of cryoglobulins the blood sample is put into a flask that is kept in a 37°C waterbath and is allowed to clot. The serum is then removed, and an aliquot is put in a refrigerator at 4°C. If cryoglobulins are present, they appear as a white precipitate that dissolves on warming to 37°C. If a cryofibrinogen is suspected, a heparinized blood sample is allowed to stand in a tube at 37°C. When the red cells have sedimented, the plasma is pipetted off, sample cooled to 4°C and observed.

peripheral circulation. In cryoglobulinemia, necrosis is not extremely rare even if transitory ischemia on cooling is more common. The fact that "vasculitis" is a common complication explains ischemic skin necrosis in other parts. Cold agglutinins usually cause general acrocyanosis including ear and nose (see Chapter 9).

9. ICHTHYOSIS

Scaling of the skin may be part of many paraneoplastic conditions. Extensive changes resembling ichthyosis are not uncommon and should be recognized, especially if they occur in an adult (143). Intense itching is a well-known complaint in Hodgkin's disease, and profuse sweating is common in patients with malignant lymphoma. van Dijk published an interesting and important review of icthyosis in 1963 (111). He describes the histories of two men, one with reticulum cell sarcoma and the other with metastasizing carcinoma of the lung. They had what he calls ichthyosiform atrophy of the skin. The number of cases with Hodgkin's disease is impressive, but other similar systematic diseases—reticulum cell sarcoma and myeloma—have been described. Only one patient with mammary carcinoma was known at the time of van Dijk's report. Congenital ichthyosis is easy to distinguish, based on the patient's history. A few patients have had a somewhat reddish skin that could not be diagnosed as exfoliative dermatitis. The desquamation sometimes was localized exclusively to the trunk. Many authors stress the point that ichthyosis appearing in elderly patients always should be suspected to be of paraneoplastic origin. From the practical point of view these dermatological symptoms may be early indications leading to diagnosis and treatment, especially in Hodgkin's disease.

We have seen a number of patients with Hodgkin's disease and ichthyosis. It is interesting that in several patients local therapy has improved their skin disease against the primary neoplasm which showed an effect. Exfoliative dermatosis has been observed to disappear after radical operation of a reticulosarcoma (503).

Flint et al have observed four patients with different types of carcinoma, all of whom had developed ichthyosis late in life (143).

10. PEMPHIGOID

Clinically, pemphigus occurring in younger adults and pemphigoid in older people are difficult to distinguish; histologically, they are quite different. In pemphigus there is an intraepidermal blister with acantholysis. Pemphigoid is characterized by bullae that develop subepidermally. A clear-cut diagnosis of pemphigoid at present is regarded to be highly suspect as paraneoplasia. Several authors regard autoimmunity as an important factor in these bullous skin diseases, and they have shown that immunofluorescence can be elicited with IgG antibody in certain skin lesions in these diseases (see also 277).

The question of immunoglobulin E and pemphigoid is discussed (214) in a recent monograph on the biological role of the immunoglobulin E, system edited by Ishizaka.

One patient with pemphigoid had a very high level of IgE in his serum. There was no cause for this finding—no known allergy, no visit to the tropics, no parasitic infection. Beutner compared the content of IgE in sera from 22 pemphigoid and 9 pemphigus patients. Seventeen of the pemphigoid patients had levels of IgE above 300 mμ/ml, whereas only two of the pemphigus patients had levels 300mμ/ml. There is obviously some overlapping, but it does not seem impossible that the pemphigus patients with unusual IgE levels were incorrectly diagnosed clinically. It appears that pemphigoid is the only skin disease—except atopic eczema—with high IgE values.

11. PORPHYRIN PRODUCTION BY TUMORS

The connection between liver tumors and the clinical picture of porphyria cutanea tarda (p.c.t.) has been much discussed. The fact that p.c.t. is connected with alcoholism and liver cirrhosis, in many instances, explains the combination of the skin disease and malignant hepatoma. Instances of this combination are not rare, but the common finding is a tumor that does not contain any porphyrins, whereas the liver tissue itself may show strong fluorescence in UV light. However, there are three very interesting instances of hepatic tumors showing intense red fluorescence whereas the rest of the liver tissue shows no signs of porphyrin production.

The most convincing example of p.c.t. as a paraneoplastic phenomenon was published by Tio et al (441).

The patient, an 80-year-old woman, developed typical dermatological signs of p.c.t. An abdominal tumor was detected, and at surgery, the tumor—a benign hepatoma—was found to be round and well-defined; the surrounding liver tissue was normal. The tumor was enucleated and carefully examined biochemically. It contained comparatively large amounts of uroporphyrin, coproporphyrin, and protoporphyrin, and separate portions of the tumor contained the same relative amounts of the three different porphyrins. It was calculated that the entire tumor contained 8.3 mg of uroporphyrin, 4.3 mg of protoporphyrin, and 3.1 mg of coproporphyrin. Macroscopically and microscopically it showed strong red fluorescence (see 33).

After the operation the dermatological and biochemical symptoms disappeared completely, and the patient remained well. Two years later she died from pyelitis. The postmortem examination did not show any liver cirrhosis or hepatoma. There was no fluorescence in UV light. Urinary and fecal porphyrins were normal.

In the medical literature there are only two reports of other similar cases. One was published by Thompson et al. A large liver tumor was detected in a 77-year-old woman who developed p.c.t. Biopsy showed a well-differentiated hepatic cell carcinoma with strong red fluorescence in UV light. The patient had no ALA nor PBG in her urine and the urinary coproporphyrin was chiefly type III, but 15% was type I. Treatment with 5-fluorouracil did not produce any remarkable effect. The patient later died. At autopsy a large liver with multiple nodules was found, and there were no signs of cirrhosis. The tumor contained rather small amounts of porphyrin on extraction, and there was retention of the porphyrins by the extracted liver tissue (434). It must be concluded that the tumor produced the porphyrins even though it was impossible to remove the hepatoma to prove that the condition was reversible (see also 33).

The second report was published by Keczkes and Barker (229).

The patient, a 60-year-old woman, had a 3-years history of p.c.t. She was treated with 5-fluorouracil through hepatic artery catheter. Her skin condition did not improve. Laparotomy was performed. A large vascular, chocolate-colored nonresectable tumor was identified in the right liver lobe; the left lobe was normal. Percutaneous liver biopsy of the tumor showed strong fluorescence in UV light.

It is interesting to speculate about the biochemical process that might explain this remarkable disease picture. The oxidation of porphyrinogens to porphyrins may be one explanation. However, we do not know how the normal porphyrinogens are kept in a reduced stage. Absence of reducing factors could also account for this porphyrin formation.

Judging from the number of clinicians who have been on a lookout for similar cases without finding any the condition is obviously extremely rare. (The report published by Waddington is not of a porphyrin-producing tumor but of an association between hepatoma and porphyria [455].) We have seen several patients with p.c.t. who have had liver cancers that did not contain any porphyrins.

12. ACANTHOSIS NIGRICANS

Even though the full-blown clinical picture is quite rare, this is one of the most classical paraneoplastic skin conditions. It was first described independently by two German dermatologists in 1890 and it is clear that there must be a large number of observations since that time that have never been published. Some years later, in 1893, the eminent French dermatologist Darier noticed that two patients had associated internal malignancy. Since that time, a large number of reviews on the condition have been published, and most likely there have been numerous unpublished observations.

The diagnosis has been somewhat obscured by the fact that although there is one very classical clinical picture that indicates the presence of a tumor, there are also similar pictures that do not indicate any neoplasia. One of these is the type of acanthosis nigricans (a.n.) that occurs in younger females who have a rare combination of extreme loss of fat and diabetes. Apparently this type is quite rare; I have seen it in only two patients. Other conditions that are similar to a.n. have been noted to occur together with some well-defined internal diseases, e.g., obesity, and should not cause too much trouble. Most authors seemingly agree that the clinical features are difficult to distinguish.

It is important to remember that the paraneoplastic a.n. usually is seen in persons over 40. However, it is found also in younger persons. The paraneoplastic form occurs equally in both sexes, but the type connected with endocrine disorders is much more common in younger females.

The adjective nigricans indicates that the skin becomes black, and these pigmentations vary in intensity. I have seen patients in whom the pigmentation dominated the picture. In others the hyperkeratosis is the most striking. A large number of verrucous and papillary formations occur in the skin chiefly in the flexures, especially the axillae, where the rough, black skin may be very striking in its appearance. The lesions are usually symmetrical; itching is present in some cases. Hyperkeratosis of the palms and soles is not uncommon. Changes in the oral cavity also have been noticed, especially on the tongue and on the upper lip, where papillomata may occur but pigmentation is rare. The development of the dermatological symptoms is usually progressive, and parallels the growth rate of the tumor.

The causal relation to the primary neoplasm is quite enigmatic even if certain facts are known. It has been stated for a long time that adenocarcinoma is the most common type of tumor, and this is certainly true. In a large collection of data by different authors it was found that adenocarcinoma of the stomach was the leading tumor. Curth found that 64% of the tumors were gastric, 27% occurred in other abdominal areas, and only 9% were extraabdominal (97). Several authors have stated that this skin lesion also may occur together with neoplasias, such as squamous cell carcinoma of the uterus and some lymphomas. Curth has stressed that adenocarcinoma is the only tumor of practical importance and pointed out that the histories of patients with other tumors have not been detailed enough to exclude a second adenocarcinoma, for instance, of the colon or the stomach. It is quite clear that the patient with a.n. should have his stomach examined very carefully.

It is of no use to discuss all the theories, or rather hypotheses, that have been constructed to explain the connection between skin and tumor. Practically all authors comment that successful extirpation of the tumor produces great improvement in the skin condition. I have analyzed a number of primary reports on this condition but have hardly found any convincing description of disappearance of the skin lesions after operation. Several papers are quoted as indicating this fact, but a closer scrutiny does not confirm this. To a large extent this seems to be connected with the fact—curious in itself when we remember the very dramatic external picture—that the tumor is usually far advanced when the diagnosis is made. It has been pointed out that there may be a long latency period between the detection of the skin lesion and the diagnosis of the cancer. To my mind this does not seem to be very remarkable since we have reason to believe that patients may carry cancers for many years before they have any distinct symptoms.

One of the few explicit statements that a.n. was modified by extirpation of the tumor seems to be the following from Brown and Winkelmann (68). Their patient was a 32-year-old woman, who had adenocarcinoma (grade II) of the rectosigmoid colon and was alive and well 14 years after the resection of the carcinoma. In a follow-up letter the patient indicated that the a.n. had regressed after surgery and was no longer evident. We also have seen a patient who had a very generalized acanthosis and a tumor of the lung. After surgery his skin symptoms were improved even though the operation was not radical. See case history E.S.

The following is an example of the diagnostic value of a.n.

E.N., a man born in 1902. The patient had been treated in 1958 for a typical "Pickwick" syndrome. He was very fat and consumed large amounts of alcohol. For about 20 years he had had a pulmonary insufficiency with retention of CO_2 and polyglobulia. He had lost 30kg in weight and his red cell levels then became normal. In 1965, he developed a rash with itching; it was thought to be dermatitis herpetiformis. During regular follow-up at the department of dermatology acanthosis nigricans in both axillae and in the inguinal region was detected in October 1970. It had not been present in March of the same year. Histological findings were classical as was the clinical picture. A differential red cell count showed 22% eosinophils. Roentgen pictures of stomach, colon, lungs, and kidneys were normal. Gastroscopy showed no tumor. His serum albumin was 3.5 g/100 ml; the haptoglobin was normal; IgA was slightly increased. In June 1971, he was readmitted for a small vascular accident with left-sided paresis. He was transferred to a hospital for chronic conditions and died in February 1972. A cancer in the sigmoid was detected at the postmortem examination.

Another important case history is the following (see color plate).

E.S., born in 1914, is a rather heavy smoker. In March 1972 he noticed a large number of small "warts" on both arms, neck, and back and some solitary ones on his face and thighs. At the same time he had pigmentations on his neck, axillae, and groin. Very extensive examination of the abdominal organs, including coeliacography and gastroscopy with biopsy, gave no findings indicating cancer. His skin lesions progressed, and in August 1974, he experienced a very marked itching all over his body. He became very heavily pigmented. A new bronchoscopy showed a possible narrowing of the bronchus going to the lingula. At operation, numerous metastases in the lymph glands around the main bronchus and in the mediastinum were found. A pulmonectomy was performed. Careful search in the lung did not produce any primary tumor, but all the lymph glands were full of squamous cell carcinoma. We saw the patient one and 3 years later. He was much improved, and his skin condition was completely normal.

For further valuable information, the reader is advised to consult the paper by Brown and Winkelmann (68).

13. PIGMENTATION AND CANCER

It should be noted that marked pigmentation of the skin may occur in carcinoma patients who do not have any signs of Cushing's disease. It seems probable that these tumors produce only real MSH and not ACTH. An interesting case of this type was published by Thivolet and Perrot (432).

A 39-year-old male developed sudden pigmentation of the face and lateral aspects of the neck following a generalized pruritus. The pigmentation had no sharp borders and was not visible on the mucous membranes. After orchiectomy for a malignant "dysembryoma" and extirpation of aortal and lumbal lymph nodes, the pigmentation faded. Six months later the pigmentation reappeared together with pulmonary metastases.

14. HYPERTRICHOSIS

For some years it has been known that patients with a "dog face" or a "monkey face" may suffer from malignant disease. Such observations were published more than 100 years ago, and a recent interesting paper by Hegedus and Schorr reviews the literature (190). Only nine cases are known so far (194). The first report was published by Turner in 1865. The patient's face and body became covered with a thick crop of short, white downy hair in 2 or 3 weeks' time. She suffered from carcinoma of the breast. In 1951, Lyell and Whittle presented a similar case with a tumor of the bladder. Of the recent seven other cases, two had a tumor in the lungs; two, in the rectum; two, in the colon; and one, in the gallbladder. The history seems to be similar in all cases. The development is rapid; the hair is fine, silky, white to slightly reddish, and is similar to nonmedullated fetal hair (see 146). The hair grows all over the body and on the ears and forehead. Scalp and sexual hair are normal. Several authors have taken biopsies, and it was noted that almost all hair follicles in these patients contain enlarged so-called mantle hair that has a special histological picture. Many of these hairs lie parallel to the epidermis. Very puzzling is the fact that Hegedus' and Scholl's patient, a 45-year-old woman with metastasizing carcinoma of the colon, also had a strawberry tongue that was typical for that commonly seen in scarlet fever.

Is this condition a relapse into fetal metabolism—like production of fetuin in liver cancer? Nothing is known about a fetal "lanugo promoting" factor.

> Chadfield and Khan (78) described a woman of 78 who had undergone nonradical surgery for a carcinoma of the rectum in February 1966. At that time no remarkable hypertrichosis was seen. In October 1966, she suffered from a trauma and was admitted with a thick growth of hair over her whole face and on her nose and ears. Similar but less dense downy hair was found in the axillae, on the trunk, and all over her arms and legs. These hairs were 20–33 mm in length. Scalp and genital hair seemed to be of the usual type. During the next few months the hair grew continuously and reached a length of 45 mm on the trunk. Histologically this hair was nonmedullated whereas the hair on the skull was normal. In February 1967, the patient died of a well-differentiated adenocarcinoma. The postmortem examination showed nothing else of special interest.

The authors state that some metabolites from the tumor may stimulate the growth of embryonic hair.

Surprisingly enough a proliferation of body hair has been seen in one special condition.

> Koblenzer and Baker described a child, who had developed leucine-sensitive hypoglycemia (240). A subtotal pancreatectomy did not cure the condition, and the patient was treated by diazoxide, 10 mg/kg/day. This produced an excellent remission. Within 6 to 8 weeks dark, silky hair involving the forehead, the sides of the face, the entire back, and the extensor surfaces of the extremities started to grow. There was no hair in the axillary or pubic areas. Biopsy specimens showed numerous "anagen" hairs. It seems evident that the hair growth was drug dependent.

In six other patients on the same drug, the extra hair was gone within a few weeks, when the drug was stopped.

It is also well-known that patients with cutaneous porphyria show abnormal growth of hair. I have seen children with porphyria congenita (Günther's disease) who had hair covering most of the forehead. The same was true of the children in Turkey who developed disturbances of porphyrin formation, like a porphyria, after being poisoned with hexachlorobenzene. Even in porphyria cutanea tarda and in porphyria variegata, moderate hypertrichosis is common Otherwise very little is known about external or internal nonhormonal factors that promote such growth (see Chapter 2, Section 5E).

15. METABOLIC DISTURBANCES (CONNECTED WITH MELANOMA)

At a very early date, Bloch, the eminent dermatologist in Switzerland, investigated the biochemistry of the skin pigment. He showed that tyrosine is changed into dopa and the latter substance into melanin by the dopa oxidase. Another pathway leading to melanin comes from dihydroxindole. It is obvious that the melanins are polymerization products. This makes their study very complicated. An excellent review of melanin biochemistry is found in Fitzpatrick et al *Dermatology in General Medicine* (142).

In 1887, Thormählen noted that patients, who had malignant metastasizing melanoma, showed a special reaction in the so-called Legal test for urinary ketone bodies: instead of developing a bright red color, the color became dark blue.

A Czech investigator had observed that patients with metastasizing melanoma excrete urine that darkens on standing. These urines contain a colorless melanogen that is oxidized into a dark brown pigment, and therefore, this condition should be called melanogenuria (368). During the last decade a number of papers have been published that have clarified many obscure points. It has been shown by several authors that the Thormählen reaction is practically specific for melanogens. The qualitative reaction is important only in patients with liver metastases of unknown origin as an easy orientation. False-positive reactions have been seen, however, and it is maintained that 1.5% of perfectly healthy people show a positive test. The test must be performed with the necessary precautions, otherwise the results are not reliable. In patients with stage 4 melanomas 83% showed a positive reaction, and the urinary concentration of indole melanogens was quite high. In 56 patients with other diagnoses, 9% were positive, and in 18 normal persons, none had a positive test. In healthy persons the oral administration of tyrosine does not produce any increase in melanogens, but in patients with generalized melanoma, who already have urinary high values, oral administration of this amino acid produced an increase.

The substances responsible for this reaction are different indole derivatives that can be identified by chromatography. In 1962, Scott from Sheffield made a study by paper chromatography in 28 patients with malignant melanoma (386). Most of the patients did not have any demonstrable melanogenuria. After extraction of the urine, the extracts were chromatographed on paper. It was then found that a band corresponding to 3,4-dehydroxyphenylalanine (dopa) was found in 22 of 28 urines. It was absent in three patients with solitary tumors and in three of the patients who clinically might have been considered to be cured. A spot corresponding to dopa-chrome probably had to do with oxidation during the procedure. In 1960, von Studnitz had reported the isolation of dopa from the urine. He studied only patients with neuroblastoma and was especially interested in dopamine. He also identified dopa in the urine from two patients with this condition. One of them had a rather high excretion, 14.7 mg/24 hr, whereas other related substances dominated the picture quantitatively. The patient excreted 2.4 mg/24 hr before surgery and no detectable amount after. The values for dopamine were exactly the same.

Since that time a large number of different indole-derived metabolites related to tyrosine (homovanillic and vanillactic acid) have been studied. In 1966, a group of investigators from St. Bartholomew's Hospital in London studied patients with malignant melanoma. These authors performed a two-dimensional chromatographic investigation of urines from these patients. Using the Ehrlich reagent, they found three different melanogens. In a number of patients with melanoma, 20% had definite melanogenuria. Treatment with cytotoxic drugs decreased the melanogen content.

Voorhess published a paper on the urinary excretion of dopa and metabolites by patients with melanoma (454). She points out that there are embryologic connections between pheochromocytoma, neuroblastoma, and melanoma since they all are of neural crest origin. Neuroblastoma belongs to the sympathetic system, synthesizing dopamine and norepinephrine. The pheochromocytoma is chromaffin and produces the two catecholamines. The metabolites of these substances are of diagnostic interest (see Chapter 10). Voorhess collected a number of urine samples from adult patients with melanoma and analyzed the amount of dopa, dopamine, and two catecholamines, homovanillic acid, and hydroxymandelic acid in each. Several patients with melanoma had high-normal values, but a few had extreme increases in dopa. Two patients also had maximal values of homovanillic acid, the same patients did not have very high dopa levels. This seems to indicate that the different metabolites vary more or less independently.

A very interesting development has been the discovery by a group in Italy that red-haired persons have a pigment in their hair that contains cysteine and dopa. One of these pigments, trikosiderin, forms polymers. Recent investigations in Lund, Sweden, have shown that a red-haired man with a melanoma on his shoulder had cysteinyldopa in his tumor. Later work by the same group (1) has shown that cysteinyldopa is present in all pigmented melanomas regardless of the type of pigment in the patient. In a group of patients with evident metastases, some patients had extremely high urinary values (up to 25–28 mg/24 hr). The large majority had values no higher than those of normal persons (363). (Normal persons rarely excrete more than 0.2 mg/24 hr, except in the summer.) Exposure to sunshine seems to increase the excretion of cysteinyldopa quite markedly. This factor should be of interest in judging results in patients (453). It is probable that quantitative determinations of this metabolite will be important in the control of melanoma patients.

16. DERMATOLOGICAL SYMPTOMS CONNECTED WITH MYELOMA, MACROGLOBULINEMIA, AND AMYLOIDOSIS

In this section, I will not discuss such symptoms as infiltration of the skin with plasma cells. These metastases are rare in myeloma, but they may be seen in the final stages of the disease. It has been shown that so-called solitary plasmocytoma of the nasopharynx has a special clinical picture with metastases found more commonly in the peripheral than in the axial skeleton. Nodular plasma cell infiltrates in the skin are also characteristic.

Some skin conditions may be regarded as truly paraneoplastic in myeloma. One is cutaneous amyloid (281). It is rare but has been described both in myeloma and in macroglobulinemia. The other two, scleromyxedema (218) and pyoderma ulcerosum (211) have a special connection with proliferation of plasma cells. Both diseases are quite uncommon. Scleromyxedema has been observed in an increasing number of patients, and now there is quite a collection of observations that these patients have an IgG M-component of moderate size in their plasma. Interestingly enough it seems that the light chains of this immunoglobulin would always be of the lambda type. (This type has a special tendency to polymerize.) There has been much discussion of whether or not this monoclonal globulin indicates the presence of true myeloma. Most observations seem to prove that it is connected with a benign monoclonal gammapathy. A number of patients have now been observed who have had the protein disturbance for many years without developing obvious myeloma. I have not been able to find any patients who later developed myeloma, and there are examples of patients with scleromyxedema who have been followed for several years (up to 9) and have not developed bone lesions. On the other hand, Feldman et al have published an interesting case history with photographic documentation of the therapeutic effects seen after Melphalan treatment (137). The skin symptoms practically disappeared both macro- and microscopically. This observation seems important since it might indicate that the primary disturbance in this disease is the formation of monoclonal globulin. We have seen two patients, who had typical *benign* monoclonal gammapathy. They were both treated with Melphalan, and the monoclonal protein fraction diminished markedly. Therefore, therapeutic effects with massive Melphalan treatment do not prove that the patient has myeloma. On the other hand, it is quite clear, that the very disabling malady scleromyxedema should be treated with Melphalan in the future.

Further follow-up on a number of patients will give information regarding the development of typical myeloma. At present it seems evident that derepression of the special immunoglobulin-forming clone producing one individual lambda IgG molecule is the basis of this condition. The fact that all these immunoglobulins have had lambda chains must be important.

17. DERMATOMYOSITIS

It seems to be established at present that the conditions polymyositis, dermatomyositis, and carcinomatous myopathy are very closely connected if not identical. On the other hand, it has been said that the percentage of patients with malignancy is much higher in dermatomyositis than in patients with pure myopathy. Williams studied this problem in 1959 and found that 92 of approximately 600 cases with dermatomyositis were associated with malignancy (495). However, combined data on 500 cases from several clinics shows only 12% with tumors. Rowland and Schneck at Columbia reviewed all cases of dermatomyositis in the Presbyterian Hospital between 1951 and 1964, and found only 3 of 48 patients had malignant tumors (369).

There seems to be a general consensus that no special tumor predilection can be observed: dermatomyositis has even been observed in a child with acute leukemia. The connection between the tumor and the dermatomyositis is obviously complicated. Patients who have improved after treatment are few, but in some of them the skin condition did not recur when the tumor did. Bluefarb has presented a case with dermatomyositis "that improved after hysterectomy for carcinoma." Another patient who had dermatomyositis for 8 years was presented by Caro. A cervical carcinoma was extirpated, followed by remission of the dermatomyositis. Some other patients have developed dermatomyositis after treatment of a tumor and spontaneous remissions are not too rare. One patient with an inoperable gastric tumor showed improvement of his myositis after laparotomy only! Such cases are of course confusing, and it seems unquestionable that the relationships between tumor, skin, and muscle are obscure.

18. BOWEN'S DISEASE

Bowen's disease was described in 1912 as a precancerous—*not* a paraneoplastic—dermatosis, occurring at times along with localized squamous neoplasms. The lesion resembles a patch of psoriasis or eczema and spreads very slowly. Graham and Helwig analyzed 155 cases of this disease and found that 25% developed a carcinoma of some internal organ within a mean time of 5.5 years (167). Other authors have noted similar but lower incidences.

2
Nervous System

One of the most difficult domains among the paraneoplasias is the neurological. It has been known for many centuries that carcinomas may have a direct effect on nervous structures. This mechanical influence, in the form of cellular infiltration or mechanical pressure, sometimes may be established only after microscopic or macroscopic anatomical examination. This is especially true of some systemic diseases, e.g., lymphomas with cellular infiltration in many organs including nerves; or collections of tumorlike cell proliferations, for instance in the brain. Macroglobulinemia is an excellent example of such a condition even if the possibility that the polyneuropathy in this disease sometimes may be caused by humoral factors can never be disproved in the individual case. Clinical symptoms of polyneuropathy—the Bing-Fog-Neel syndrome—are not uncommon in severe macroglobulinemia and will be treated later (44).

It was not until Denny-Brown, in 1948, published his paper on primary Sensory Neuropathy with Muscular Changes Associated with Carcinoma, that a "toxic" effect was discussed (109). This publication started a wave of observations on similar cases. Russel Brain and colleagues in London followed this lead and collected a large number of similar observations on lesions also in other parts of the nervous system. The different paraneoplastic disturbances in the function of the nervous system have now become very numerous.

1. GENERAL DISTURBANCE OF BRAIN FUNCTION

The most puzzling part in these syndromes is played by tumors that produce such humoral changes that there is marked disturbance of brain function, i.e., an encephalopathy. The classical and most common is hypercalcemia which will be discussed in a special chapter. The severe symptoms in this condition are chiefly cerebral, and it is not uncommon to find patients with paraneoplastic hypercalcemia admitted to a hospital because of mental symptoms. This condition is a paradigm of a true paraneoplasia. Treatment of the tumor reduces the calcium level, and the neurological symptoms may disappear completely. Also, such objective findings as electroencephalograms are often much disturbed before treatment and become normal after (130).

Patients with adrenal carcinoma may have the psychological disturbances that are seen in hyperadrenocorticism, insuloma patients may have severe hypoglycemia, and many patients with lung tumors have hyponatremia with epileptic fits or coma. The last condition and ectopic formation of ACTH with hyperadrenocorticism are truly ectopic paraneoplatic phenomena.

2. HYPERVISCOSITY

In 1956 Wuhrmann published a paper about "coma paraproteinaemicum." He pointed out that clinical conditions with disturbed consciousness, ending as coma, may be seen in patients with increased levels of myeloma globulins or macroglobulins (505). I have pointed out that several of our comatose patients with myeloma had hypercalcemia—one of the common complications of myeloma that causes severe mental disturbances. In 1960 and 1963 Fahey and his colleagues (407) published their experiments with massive plasmapheresis in patients with hyperviscosity from severe macroglobulinemia. They were able to show that also general cerebral symptoms disappeared when the viscosity of the blood was brought down to normal or more normal levels. These investigations proved that the hyperproteinemic coma really exists under rare conditions. It should be remembered that hyperviscosity is quite rare in myeloma, whereas hypercalcemia is fairly common. The reverse is true of macroglobulinemia.

To recognize the hyperviscosity syndrome it is necessary to observe some clinical features. In my first publication on macroglobulinemia, I pointed out the fact that there was an unusual fundoscopic picture in some patients. The veins were greatly engorged, and there were numerous bleedings in the eyegrounds. A generalized bleeding tendency, e.g., from gums and nose, as well as a prolonged bleeding time were also noted. Measurements of serum viscosity showed that this was much increased and that there was an unusual relative increase at lower temperatures (449). This hyperviscosity syndrome has been studied in great detail by several groups during the last years, and it is clear that patients with very high macroglobulin levels usually have signs of hyperviscosity. When the levels are moderately increased, the syndrome is very rare.

The first step in the diagnosis of this condition is a good history of bleedings. Women may bleed more or less continuously from the gums without having increased blood loss from the uterus, or they may have profuse menstrual bleedings. Examination of eyegrounds is very important. Papilloedema may be found without any indications of localized neoplasia intracranially. Sometimes the patients complain of impaired vision. It is important to remember that the high viscosity causes very marked sludging of the blood and it is probable that this is one of the causes of the impaired vision. Slit lamp examination of the conjunctival vessels may show granular streaming, similar to that seen in any patient with a maximal sedimentation rate.

The next important diagnostic step is the determination of plasma viscosity with a simple Ostwald viscosimeter. This is a simple procedure that gives objective measure of the degree of the condition.

In a few patients headaches have been relieved by plasmapheresis (407). I have seen several comatose patients, but I have not been able to relieve this symptom except in one patient with macroglobulinemia. The fact that so many patients belong to the higher age groups makes it difficult to decide if symptoms such as seizures and strokes are connected with hyperviscosity. They may be caused by infiltrates of lymphatic cells in the dura or in the brain itself causing symptoms of cerebral tumor. I have seen both kinds. Atherosclerotic lesions are probably the most common cause of such central nervous symptoms. It is difficult to tell if the progressive deafness that is noticed in many of these patients is related to the viscosity. I am inclined to regard this as one aspect of the peripheral neuropathy that is so common in these patients. This syndrome (Bing-Fog-Neel) has nothing to do with viscosity, and I regard it as a special type of carcinomatous neuropathy. Solomon, who has studied the hyperviscosity syndrome in great detail, before and after extensive plasmapheresis, is of the same opinion. He stresses another interesting syndrome that improves after plasmapheresis—postural hypotension. Two of his patients with very high serum viscosities before treatment regained normal standing blood pressures after plasmapheresis. Solomon showed that the EEG returned to normal after plasmapheresis in two patients. In one it returned to the previous pathologic pattern, when serum viscosity increased (405).

Several observations have shown that focal symptoms of different kinds may be caused by localized accumulations of pathological cells in the dura and the brain. In such cases local irradiation may be quite effective in relieving symptoms. Such cases are rare, however, and in my own experience I know of only one patient who developed intracranial lesions after having been ill for at least 18 years. Plasmapheresis therefore should not be started unless the diagnosis of a high viscosity syndrome is clear.

3. LEUKOENCEPHALOPATHY

This disorder has been called progressive multifocal leukoencephalopathy (PML) and has been studied in great detail by the group at Massachusetts General Hospital in Boston. In 1958 Åström, Mancall, and Richardson published a classical paper on three cases with this condition. Two patients had chronic lymphocytic leukemia and one had Hodgkin's disease. It was therefore suggested that this may be a complication of lymphoma (10). Later many similar cases were published from different centers.

The disease is obviously not very rare since it occurs in many parts of the world. More than 60 cases have been reported, and it is now clear that a disease picture resembling PML is most common in the two diseases already mentioned, but also occurs in lymphosarcoma, myelocytic leukemia, polycythemia, and myeloma. It has been observed in only three patients with carcinoma, and also in two patients treated with immunolytic drugs. (One had SLE and the other, renal transplanation.)

In 1961 the possibility of a viral etiology was discussed and later tested by attempts to cultivate an organism from brain tissue (352). In 1965 two teams working with electron microscopy demonstrated particles that seemed to be papova virions. Papova is a group of DNA viruses. Human warts may be caused by a closely related papilloma virus. The virions are very numerous in the nuclei of oligodendrocytes. In 1971 workers in Madison were able to isolate a virus strain—called J. C.—that they regarded to be a new member of the papova group and it is now quite clear that different viruses from this group can be grown on suitable tissue culture media (see 76 and 332).

This virus is found in PML and the viral etiology of this condition is now well established (311). Therefore, it should not be included among the truly paraneoplastic conditions even though it is clear that the virus grows only in the brains of patients with impaired antibody production, either primary or secondary to cytostatic drugs or both.

We observed a man whose diagnosis in 1962 was lymphatic leukemia. His total gammaglobulin value at that time was 0.5 g/100 ml (low). In 1963 prednisone was started and in 1966–1967 he had three courses of cyclophosphamide. This caused trombocytopenia and had to be stopped. During 1967 the spleen became much enlarged, and in October of that year he suffered a rapidly progressing bilateral loss of vision (increasing central scotoma) with dizziness developing in a month's time. The CSF contained 10 mononuclear cells per microliter and protein (61 mg/100 ml). The electrophoretic picture of the CSF was normal. The patient became lethargic, developed hemiparesis, and died in February 1968, after being comatose for several weeks. The histology was quite typical (347).

During the last 5 years we have seen six hospitalized cases in a population of 250,000 (Rausing, personal communication).

The clinical picture seems to be fairly typical. Dr. Rausing's Department of Pathology found the characteristic electron-microscopic picture in one patient, who was born in 1875.

In 1959 she had an M-component of 3 g/100 ml with a serum albumin of 4.0 g/100 ml. She also had small lytic foci in her skull. In 1960 her M-component had increased to 3.5 g/100 ml in August and 3.9 g/100 ml in November. A sternal puncture showed an increase in plasma cells. Alkeran treatment was started according to our usual schedule. In October 1961 the M-component had decreased to 2.5 g/100 ml, and continued to decrease to 0.6 g/100 ml in March 1964. The serum albumin level remained normal. She was in very good shape all the time despite her age. Her background gamma globulin increased to normal values, and her myeloma globulin, an IgG kappa, decreased to 0.4 g/100 ml. It became practically invisible in 1971, when she was transferred to a hospital for chronic diseases. Treatment was continued there. Her last sternal puncture was performed in September 1972 and did not show any increase in plasma cells. She was not anemic and had no leukopenia but it was thought unnecessary to continue Alkeran therapy. In November 1972 she developed an epileptic fit with a transitory right-sided hemiplegia and died on July 6th 1973 at the age of 98! The anatomical diagnosis was multiple myeloma with small collections of plasma cells. Macroscopically, her brain showed numerous small greyish, gelatinous foci in the cortex. In the white matter there were very widespread lesions that could not be observed macroscopically. Electron microscopic examination of the grey and the white matter showed marked changes in the nuclei of the glia cells. Typical virions were seen, partly spherical and partly tubular. The diagnosis was without any doubt leukoencephalopathy.

4. CEREBELLAR SYMPTOMS

Carcinomatous cerebellar degeneration is a relatively rare disorder. Brain and Wilkinson reviewed their findings in 1965 and found 17 cases. The youngest patient was 36 and oldest, 70 (60). Nine patients had lung carcinoma of different types, five had ovarian carcinomas, and only two had mammary carcinoma (The number of patients with ovarian carcinoma is surprisingly high.).

The onset may be very rapid with unsteadiness and difficulties in walking. Nearly all of Lord Brain's cases were ataxic in both the upper and lower limbs. Several patients were unable to sit up in bed. Fourteen of 17 had marked dysarthria, some were difficult to understand. Nystagmus was rather uncommon. Disturbances in the function of cranial nerves were common, whereas only a minority had weakness of the limbs. Pain occurred in three patients and deafness, in two. Thus, it is clear that the cerebellar symptoms usually dominate the picture, even though they may be combined with other neurological signs seen in carcinoma patients (59).

Brain stresses the fact that the connections between the carcinoma and the lesions in the cerebellum are difficult to analyze. He observed one patient who did not develop signs of cerebellar dysfunction until some months after the removal of the tumor. She survived for 6 years. The type of tumor she had is not stated and nothing is mentioned about metastases being found after death. It is not impossible, however, that a person may live for 6 years with metastases. The pathology of the condition is very interesting since it shows an almost specific loss of the Purkinje cells. Brain examined 6 patients at autopsy, and found patchy degeneration of the dentate nuclei as well as degeneration of the ganglion cells in the ocular motor nuclei and in the hypoglossus. In some patients there was degeneration of the anterior horn cells of the spinal cord.

We have observed a typical case in a patient with lymphatic leukemia.

This was a man born in 1894. In the summer of 1969, the diagnosis of cerebellar ataxia was made. He had an enlarged spleen, anemia, and signs of malignant lymphatic disease. He had severe bilateral ataxia in his arms and legs, but he did not experience vertigo. There was evident dysarthria. No paresis, no sensory disturbances, and no signs of pyramidal involvement tracts were found. In the right supraclavicular fossa as well as in the left axilla, two enlarged lymph glands were found. The sternal marrow showed cells that were interpreted as being reticulum cells. His WBC count was only 3000/cu mm; his platelet count was 75,000/cu mm. He had a slight hypochromic anemia and an ESR of 10 mm in one hour. Electrophoresis of the serum showed nothing pathological. A splenectomy was performed on the indication hypersplenism. The pathology was regarded as lymphoma of "lymphocytic type," "immunocytoma," "lymphoplasmacytoid type (Lukes)." After the surgery his platelet and WBC counts increased considerably. The patient had slight diabetes, but otherwise the cerebellar symptoms still dominated the picture. His spleen and his blood were examined in the hope of establishing some viral etiology. Nothing was found, however, except a high titer of herpes simplex (1/128). The tests were repeated and the titers were slightly lower, 1/64 in one sample and 1/32 in another. During the next year severe progression of the cerebellar syndrome occurred. He was admitted for the last time in April 1970, cyanotic and pulseless. The next day he developed asystole and died. At the postmortem examination, the bone marrow was found to contain well-defined, small lymphocytic infiltrates. The lymph glands had a completely normal architecture and showed no signs of a malignant lymphoma. It was evident that the lymphoma was chiefly located in the spleen. Examination of the brain showed status lacunaris with numerous encephalomaciae. Certain parts of the cerebellum were completely normal with well-preserved Purkinje cells. In other parts long stretches of these cells were absent. A few cells seemed to be damaged. There was no inflammatory reaction.

In recent years much has been published on scrapie, a neurological disease in sheep. The damage to the nervous system occurs predominantly in the cerebellum, where there is gross outfall of Purkinje cells. Lesions of the hypothalamoneurohypophysial system are also seen. These probably lead to different metabolic disorders such as diabetes insipidus and enlargement of adrenal glands. The etiology of the disease had earlier been obscure, and it was maintained that heredity was an important factor, and that the mode of inheritance may be recessive. This is interesting since it has been definitely proved that scrapie is caused by a "slow" virus, and that there are many parallels with Kuru which is found in some Papuan tribes in New Guinea. The mode of infection with Scrapie is not known, but it seems well established that head-hunting is connected with the spread of Kuru. Cannibalism, centered on the enemy's brain, is an important rite in these tribes, and some good evidence has been presented that the infectious agent comes from ingestion of infected human brain. These two diseases are mentioned only to demonstrate a viral etiology of such chronic neurological conditions. It may be a parallel to the story of progressive multifocal leucoencephalopathy.

5. PERIPHERAL SYMPTOMS

A. Myoneuropathy and Myasthenialike Conditions

The most common among these syndromes is peripheral neuropathy. This may be purely sensory or combined with motor symptoms. Sometimes it may be difficult to distinguish the neuronal and the muscular disturbances. There is also a well-known paraneoplastic dysfunction of the neuromuscular junction itself with a picture resembling myasthenia but different from true myasthenia gravis.

As already mentioned, most authors seem to agree that it may be difficult to distinguish between primary disturbances of the muscular system and the motor innervation in a number of paraneoplastic syndromes. Tyler has recently tried to avoid this point by writing a chapter on paraneoplastic "syndromes of nerve, muscle and neuromuscular junction." He points out that polymyositis occurs together with carcinoma, and it is clear that a number of cancer patients must have a primary disturbance of muscle function (447). This problem has been investigated by Belgian authors, who found that changes in the nerves were not too uncommon in cases with disturbances clinically regarded as primarily located in the muscles (196). Progressive wasting and weakness of muscles also may clearly be the result of tumor cachexia. It has been pointed out too that these symptoms are more marked in proximal than in distal muscles. Tenderness over the muscles may be quite marked, but reflexes are not absent. This more or less polymyositic picture may be combined with skin symptoms (see Chapter 1, Section 17).

Dermatomyositis for many years has been regarded as probably paraneoplastic even though the proof—complete reversal after operation—is still lacking. It seems clear, however, that the term used by different authors: "carcinomatous neuromyopathy" shows the difficulties and the prefix "dermato-" may be added in order to make the picture still more confused.

Classical dermatomyositis is connected with proximal wasting of muscles that may become very severe. Atrophy of neck muscles may cause difficulties in keeping the head straight, and paresis of the trunk is sometimes so severe that it causes respiratory distress. The dermatogical symptoms consist of erythema, typically around the eyes and periorally. Small red patches may occur on the hands and fingers and a more widespread erythema on limbs and trunk. Williams made a very complete study of dermatomyositis from his own experience and from the literature.

Wong published a paper on dermatomyositis and squamous cell carcinoma of the nasopharynx. He mentioned that 23 cases of dermatomyositis in the Chinese had been published. Of the patients over 40, 70% had an internal malignancy; 75% of these cancers were nasopharyngeal carcinomas, a disease that is especially common in the Chinese.

We have seen a number of patients with dermatomyositis but only very rarely was it combined with a carcinoma. It is said that the muscle disease occurs much more commonly in patients with carcinoma of the lungs, and this is probably true (94). The nosologic problem is complicated by the fact that "myastheniaIike" conditions seem to have more definitely polymyositic pictures i.e., the type of myasthenia that has been studied in detail by Lambert, Eaton, and Rooke. It is quite clear that they differ from myasthenia gravis in many respects.

In an extensive monograph by Walton and Adams the problem of polymyositis is treated in great detail (481a). At the time (1958), very little seems to have been known about the connection between polymyositis and malignant disease, even though the subject is treated and a number of case histories are quoted. Several of the cases described as proximal motor neuropathy in bronchogenic carcinoma in reality may have been myopathy. It is quite remarkable that only one example of cure after radical treatment of the tumor is quoted.

This patient was a 55-year-old man, who was seen by Adams. For 5 months he had had pulmonary symptoms accompanied by exfoliative dermatitis, dysphagia, and severe muscle weakness involving the shoulder, pelvic girdle, and trunk. A lung tumor was resected and soon after all symptoms of dermatomyositis disappeared. He was completely well at follow-up 3 years after the resection.

B. Myasthenia

In most published reports thymoma is the only tumor that is commonly associated with myasthenia, and it is believed that true myasthenia occurring with other carcinomas is no more common than might be expected by chance. Some cases of oat-cell carcinoma with classic myasthenia have been published, but in view of the recent ideas on similarities or identity between oat-cell carcinoma and thymoma, it may well be that the so-called oat-cell tumor was actually a "thymoma" (333). It may be said that classical myasthenia is probably not related to any tumor other than the thymoma. The links between the two conditions are very obscure, but recent findings of antibodies to the neuromuscular junction in thymoma may give a clue.

C. Lambert-Eaton Syndrome

This rare syndrome (246) named after its principal investigators, occurs almost exclusively in small-cell pulmonary carcinoma and resembles myasthenia gravis. However, classical myasthenia is much more common in young women and is evident chiefly in the ocular-bulbar muscles. The decamethonium test shows supranormal sensitivity, and neostigmine is a very effective therapy. In the Lambert-Eaton syndrome the reverse is true, and there are also electromyographic differences. The connection between the Lambert-Eaton syndrome and tumors is quite obscure, and no reports have been published of the disappearance of the neurologic symptoms after surgery. "Striking remissions" (without stating any details) have been reported, however.

The clinical findings in this syndrome are weakness and easy fatigability of proximal muscles, especially in the pelvic girdle and thighs. The principal feature of the defect in neuromuscular transmission is a marked block in rested muscle but a marked facilitation in the active. Of 30 patients seen by Lambert and Rooke (239), 22 subsequently developed intrathoracic tumors, 19 of which were bronchogenic. Eight of the 30 patients showed no signs of a tumor, but none of these was examined post mortem. It is quite clear that the condition is rare as authors with some experience always have had patients sent to them from many different hospitals. Lambert and Rooke saw subjective improvement in two patients after treatment. One, a 51-year-old man, showed some improvement in the third postoperative week. Eleven months after surgery he had no EMG disturbance of transmission, and this remission lasted another 15 months, after which both subjective and objective symptoms reappeared. The ultimate fate of this patient has not been published, but it is probable that he suffered a relapse.

Norris et al. have described their observations of a man, who had the typical Lambert-Eaton syndrome combined with hyponatremia and hypotensive episodes. Supplementary salt and corticosteroids corrected these symptoms but not the neuromuscular disorder. He could not sit up nor climb stairs. His EMG was typical. He had a lung mass that was irradiated. The lung tumor disappeared and his strength improved dramatically. He was able to return to work and his EMG was much improved. After 4 months his EMG again showed abnormalities, and X-ray films demonstrated tumor growth. Death disclosed a small oat-cell carcinoma with metastases in the adrenal glands.

The abbreviation SLE (S for syndrome) is ambiguous and should never be used as SLE usually means systemic lupus erythematous.

D. Polyneuropathy and Cancer

The problem of polyneuropathy and cancer has been discussed extensively in recent years and opinions are rather varied. Neurologists at the Massachusetts General Hospital are of the opinion that this is a rather rare condition, and Victor only found 25 well-described cases in the literature (1959), the large majority occurring together with carcinoma of the lung. British neurologists at the London Hospital (58, 59) have collected considerable data and are of the opinion that it occurs with carcinomas in different sites. The question of the etiology of polyneuropathy is difficult. It is probable that the French author

Auché in 1890 was the first to point out that a metabolic disturbance and not metastatic infiltration may be responsible. A number of later authors, chiefly from Germany, talked about toxins. In 1948, Denny-Brown described two cases with degeneration in the ganglia of the posterior root and sensory neuropathy. The etiology of this polyneuropathy is a difficult question. It is well known that polyradiculitis of the Guillain-Barré type is often caused by an infection. But the connection between malignant neoplasms and this type of neurological disease only recently has been well established. The group at the London Hospital has collected their patients with peripheral neuropathy and carcinoma and have stressed that the systematic search for occult carcinoma in older patients with unexplained peripheral neuropathy is very important. Of the 31 cases they have seen, 24 have had cancers in the lung, three in the stomach, two in the breast, two in the thymus, 1 in the colon and 1 in the pancreas. The incidence was the same in men and women. Cancer in other sites also has been observed in isolated cases.

The onset is most commonly subacute, but there are also examples of a more insidious or a more rapidly progressive type. In many patients the clinical picture is indistinguishable from that of the Guillain-Barré syndrome. Spontaneous remissions have been seen in patients with progression of their neoplasm. (This only applies to patients with peripheral neuropathy; ganglioradiculitis is much more serious.) The group at the London Hospital are very skeptical about the improvement after treatment of a cancer, but it is true "that early detection of symptoms of carcinoma from neurological findings has saved several lives in their series." It is remarkable that the authors do not describe any patients with successful therapeutic results from operation but mention that this happened in "some of our patients."

It has been maintained that about 2% of patients with Hodgkin's disease have peripheral neuropathy. At the London Hospital 1.4% of the patients with Hodgkin's disease had peripheral neuropathy, 3.2% with myeloma and, 1.5% with polycythemia vera. One patient with Hodgkin's disease had three episodes of remitting peripheral neuropathy over a 6-year period before Hodgkin's disease was discovered. At autopsy this patient was found to have had also a clinically silent cortical cerebellar degeneration. Symptoms of peripheral neuropathy also have been seen in the reticuloses.

Of interest are the results from a study of paraneoplastic neurological symptoms in an unselected consecutive group of patients with bronchogenic carcinoma (303). Morton, Itabashi, and Grimes collected data on no less than 23 cases of neuromyopathy from the University of California Medical Center and the Veterans Administration Hospital in San Francisco by studying 175 charts of randomly selected patients. In 22 cases the diagnosis of the neurological condition was confirmed by careful postmortem examination and in one by clinical follow-up. The authors also studied 142 previously published cases. The results were the following: 55% had carcinomatous myopathy, 25% had polyneuropathy, and 13% had pure sensory neuropathy. The percentages are almost identical to those in the literature. Subacute cerebellar degeneration and encephalomyelopathy were much more common in the San Francisco study. These conditions occurred in about 40% of the cases, whereas the percentages in the literature were 10 and 15%. The paper gives excellent illustrations and case histories. The majority of the 23 cases were oat-cell carcinomas as were those cited in the literature, but the most common type of pulmonary carcinoma among *all* patients with pulmonary carcinoma in the material was squamous cell.

In the discussion of this paper Grimes, who is one of the authors, mentions an interesting case.

> The patient, who had a big infiltrate in the left upper lobe, became mentally disturbed and forgetful and developed a blurred speech. He also had peculiar athetoid movements of the face with spasmodic twitchings and generalized muscle weakness. All other studies including a brain scan were negative. His left lung was resected, and within 3 months after operation, most of the neuropathological symptoms disappeared. The muscle weakness improved, and 4 months after surgery, he was able to return to full-time activity at his office.

It is well worth noting that these patients usually have symptoms of several neurological disturbances at the same time (197). (For additional case histories and illustrations, the interested reader is referred to the original article [303].)

The literature contains several observations that have been interpreted as indicating a real connection between amyotrophic lateral sclerosis (ALS) and carcinoma. The most convincing data were published by Norris and Engel (55). These authors are very careful in their conclusions and only express the hope "that early detection and successful treatment of a remediable neoplasm will alter the downhill course characteristic of ALS." A large number of different neoplasms have been found together with this neurological disease. A recent contribution to this discussion was made by Bauer et al (26) who observed a patient suffering from signs of motor neuron disease (ALS) since 1971. This patient had had an unexplained maximal ESR since 1969. The patient died from paralyses in 1973. During his last hospitalization it was found that he suffered from macroglobulinemia with an IgM kappa of 23g/L. A very careful postmortem examination confirmed both diagnoses.

E. Neuropathies Connected with Myeloma, Amyloidosis, and Macroglobulinemia

Myeloma

Of special interest is a syndrome that has been much discussed among specialists in myeloma. The first observations regarded only the combination of neuropathy and myeloma. Already in 1958, Victor et al from the Massachusetts General Hospital published a report of 5 cases of their own who had polyneuropathy that was not caused by compression of the nerves nor by amyloid (451). (The combination myeloma with amyloidosis and amyloid in the nerves has long been known, and this may also be regarded as a paraneoplasia.) Victor's patients ranged in age from 45 to 58 and seemed to be rather young to have a myeloma. In several radiologic studies, sternal puncture did not reveal any obvious signs of myeloma. Two patients had a very long disease course—one of 9 years—and both had no progression of the myeloma even though the neuropathy was progressing very slowly. Two patients died from the polyneuropathy and, the diagnosis myeloma was not suspected until late in the history. Victor mentions that one of the cases had osteosclerotic lesions (see also 2). (I have stressed this point in several publications and collected some personal observations and cases from the literature indicating that a relationship must exist between these two symptoms [466].) It was also evident that the majority of the patients were much younger than we see in a representative material of myeloma cases. The number of plasma cells in the bone marrow was low. The myeloma globulin in the serum was not much increased, but the protein in the CSF was elevated. Since Victor's publication, several new observations regarding myeloma and polyneuropathy, with or without osteosclerosis, have been published (103, 301).

During the last 5 years, Japanese authors have shown that this unusual triad may be still more complicated. They have published reports of five men and one woman, all of whom were below the age of 51. These patients showed only a very slight increase in plasma cells in the bone marrow. All had polyneuropathy, pigmentation, thickening of the skin, peripheral edema, and excessive perspiration (210). None had Bence Jones proteinuria. All five males had gynecomastia, three had noted impotence, and three had sclerolytic bone lesions. Four patients had diabetes mellitus.

This very bizarre syndrome seems difficult to explain but observations on some other patients seem to indicate that several of these symptoms must be regarded as paraneoplastic.

We recently published reports on two patients from Malmo who had similar symptoms (467), One was a man aged 50, the other, a woman of 48. Both had the same type of myeloma and a Guillain-Barré-like syndrome. The man perspired excessively, was generally very thirsty, and also had a transitory glucosuria. He died from paralyses. The other patient developed very marked hypertrichosis on her face, and arms especially on her legs, where a tuft of long black hairs covered the tibial tuberosity. When the patient was treated actively with Alkeran, her hypertrichosis disappeared. Her neurological symptoms improved slowly; she is alive 9 years after the first signs of plasmocytoma and is able to do her household chores. This remarkable but very slow improvement of the neuropathy together with remission of the plasmocytoma seems to indicate that the polyneuropathy may be secondary to the proliferation of plasma cells.

Are there any other known cases of remission or cure of the neuropathy when the myeloma was treated?

In 1972 Davis and Drachman published reports on two patients at the Johns Hopkins Hospital. Both were men, one 46 years old and the other, 49. Both had a neuropathy (104). One man had a multiloculated cystic lesion in the right acromion but no other lesions. Immunoelectrophoresis of his serum showed a monoclonal Ig. Results of biopsies from sternum and the ileum were normal; a tumor specimen showed plasma cells. The patient received local irradiation (4000 R). Shortly after, the burning sensations in hands and feet that he had experienced for 2 years before treatment disappeared gradually. After 6 months he could walk unaided. Two years later there was no recurrence of the plasmocytoma, and the muscular power in his upper limbs was normal, except in the small muscles of his hand. The M-component of his serum globulin was still present. The patient was able to return to work. At reexamination 10 years later, he was still in good shape with no recurrence of the plasmocytoma.

The other patient had had polyneuropathic symptoms for 2 years with bilateral foot drop and impotence. His CSF protein was 225 mg/100ml. Treatment with large doses of steroids had no effect. A radiolucent focus with sclerotic margins was found in his left humerus. Biopsy of the bone tumor showed a characteristic plasmocytoma. Results of a sternal biopsy were normal. His serum level was increased. Local irradiation (3750 R) produced rapid improvement, and after 9 months only mild distal weakness remained, potency returned, and his serum immunoelectrophoresis was normal.

Davis and Drachman tabulated the results of therapy in 30 cases of myeloma with polyneuropathy. No patient given only steroids was improved. Three of five patients improved with radiotherapy; one of 11, with antimetabolite therapy and one of 4 without treatment. Seventy-five percent of the patients were males, 24% had solitary lesions, 55% had osteosclerotic infiltrates. Amyloid was not present. The authors believe that polyneuropathy is a remote effect of the myeloma, and they discuss the possibility that immunologic mechanisms may be at work.

Amyloid

The combination of amyloid and myeloma is complicated and may lead to symptoms of neuropathy. This is of different types. In some patients there is infiltration of nerves with amyloid resembling the situation in the so-called hereditary (Portuguese) amyloidosis that has also been found in Japan and in Sweden. In other patients there are symptoms of the carpal tunnel syndrome. Amyloid masses collect in the carpal tunnel and exert pressure on the median nerve. In our experience this is a rare phenomenon, but we have seen it in two patients, both of whom were much relieved after surgery (466).

The following case history illustrates the complicated mechanism that may be at work.

The patient was a man born in 1891. In 1958 he started to have shoulder pains with irradiation into the arms and fingers. He also noticed decreased muscular power in the fingers with paresthesiae. He had been treated with physical therapy and ACTH without any real improvement. The condition progressed very markedly, and in July 1965 most of his musculature was atrophic. He experienced tingling in his fingers, and on cooling, the fingers became bluish. His ESR had been very high (130 mm/1 hr), and serum electrophoresis showed a low albumin level and a high level of monoclonal type gamma globulin. Wasserman reaction could not be performed because of anticomplementary effect. The patient was quite emaciated but was able to walk. A film of his lungs showed very numerous miliary circumscribed densities in both lungs. Skeletal films disclosed widespread myeloma with typical foci in the skull. The patient was admitted to the department of medicine in Malmo on August 19, 1965. He had a typical macroglossia and a very marked atrophy of his entire musculature. He also showed signs of a generalized malabsorption with pathological xylose and Schilling tests. His sternal marrow showed 6.5% plasma cells. His serum contained a very massive M-component of 3.6 g/100 ml that was Ig-GL. His urine contained a small amount of albumin but no Bence Jones protein. His serum albumin level was low (2.0–3.7 g/100 ml). Hematologically he had slight anemia; 11.2 g/100 ml, (70%), RBC: 3,400,000/cu mm, WBC: 7900/cu mm, platelets: 180,000 cu mm. His creatinine level was 1.4 mg/100 ml rising to 1.8 mg/100 ml. and his BUN level was 53 rising to 120 mg/100 ml. The diagnosis of myeloma with possible amyloidosis was made. At the postmortem examination amyloid was found in the muscles of his tongue, and there were very massive infiltrations of amyloid in the walls of many vessels, especially in the lungs, intestine, and myocardium. The kidneys did not show much amyloid, however. The muscles showed a picture compatible with the diagnosis neurogenic atrophy.

Macroglobulinemia

This may cause neurological symptoms through three different mechanisms. Most easily explained are the symptoms provoked by infiltration with pathological cells. Such infiltrates may be seen in the peripheral nerves, in the dura producing pressure on the brain and signs simulating a cerebral tumor. In a few cases tumorlike collections of these cells have been noted in the brain with a true picture of cerebral tumor. These symptoms are not paraneoplastic. In other cases, there is a special kind of polyneuropathy. These patients suffer from a slowly increasing but usually not invalidating numbness and tingling in their fingers and toes for many years. Several patients also complain of increasing deafness. In our experience this symptom complex is not extremely rare. It was first described together with "dysproteinemia" by the Danish authors Bing, Fog and Neel, who have given their name to the syndrome (44).

Solomon analyzed the neurological symptoms of 27 patients with macroglobulinemia from the National Cancer Institute (405). He found only five with peripheral neuropathy. One also had amyloidosis. Bigner et al have described a patient who had a sensory motor neuropathy for 2.5 years (42). She was treated with chlorambucil, prednisone, cyclophosphamide, and Melphalan, but she did not respond. Nerve biopsy showed amyloid fibrils. (This patient probably had an unusual amount of 7S, i.e., low molecular weight IgM.) The authors do not believe that amyloid caused the symptoms but discuss the possibility of a remote effect of cancer (see also 249 and 318).

To me it seems probable that the mechanism producing the neurological symptoms is the same in macroglobulinemia and in multiple myeloma. Logothetis has collected 182 observations on patients with macroglobulinemia and neurological symptoms (267). Fourteen patients had peripheral neuropathy. In three the clinical picture was of the Guillain-Barré type, six had spinal type, and five had bulbar-type neuritis (progressive deafness). A large number of patients had more or less generalized symptoms. Solomon has been especially interested in the hyperviscosity syndrome and our own experience is very similar: stupor, coma, ataxia, vertigo, and so forth may improve dramatically after plasmapheresis, whereas the hyperviscosity does not seem to be of importance for the development of polyneuropathy (405). Neuropathic joint disease (Charcot) also has been described (387).

In a recent publication from Göttingen, Iwashita et al described a man with macroglobulinemia and polyneuropathy (216). His CSF was normal. A biopsy specimen from the nervus suralis showed that the nerve was thickened. The nerve contained amorphous masses interstitially in the peri- and endoneurium. These gave strongly positive reactions with PAS, and the authors believe that they contained IgM. (It is not impossible that the precipitate may consist of parts of the macroglobulin molecule.) After cytostatic treatment with chlorambucil his condition improved steadily but he developed a carpal tunnel syndrome. Beautiful pictures illustrate the anatomy.

3
Blood

1. Blood Morphology and Tumors
2. Disturbances of Blood Coagulation in Cancer

1. BLOOD MORPHOLOGY AND TUMORS

A. Anemia

The most important hematological symptom in cancer is of course anemia. This has been known to be a common effect of cancer for a very long time, and many methods of treatment have been tried with meager results. It is regrettable that our knowledge about the relationship between the tumor and the depressed function of the erythropoetic apparatus should be so scarce. It is very difficult to find publications giving a well-analyzed and extensive study of tumor patients *without external blood loss* who have been treated only by extirpation of the tumor and possibly with a few transfusions during the operation and who have had their postoperative blood values checked for some time after the extirpation. Such critically collected data are sorely needed. I suspect that most authors take it for granted that a normochromic anemia—usually with a lowered serum iron level combined with a rather low transferrin level—will be corrected after extirpation of the tumor. Proofs are difficult to find, however, since most patients have been treated with different preparations.

Banerjee and Narang conducted an excellent study of general hematological changes in malignancy (19). They studied 65 cases of malignant disease in different organs, the patients were not receiving radiotherapy or chemotherapy. Ninety per cent of the women and 78% of the men showed normocytic, hypochromic, or normochromic anemia. However, it is important to notice that 15% of the men did not show any significant anemia in spite of widespread metastases. Leukoerythroblastic anemia was seen in four patients, and malignant cells were found in the marrow from all four patients. Significant leukocytosis (WBC over 10,000/cu mm) was found in 23% of the cases with varied localization of the tumor. It was rather remarkable that about 12% of all the patients had leukopenia accompanied by moderate to severe anemia. Thrombocytopenia was observed in 11% of the patients. Serum iron concentration was not determined in all patients and TIBC was not determined in any. Ferrokinetic studies were performed. The iron clearance rate from the plasma was decreased and iron utilization by the bone marrow was poor. A very interesting finding was a significant increase in plasma volume in 50% of the cases. This is very important since it may cause spurious anemia by dilution of the red cells. Determinations of total haemoglobin should have been of interest.

An excellent study of the reversibility of anemia after removal of a tumor was made by Greenberg and Creger (170).

The patient, a 69-year-old woman, was treated at Stanford University Hospital. She had lost 40 pounds, had anorexia and fatigability, and was admitted because of iron refractory anemia. On admission she was cachectic with moderate glossitis. No increase in size of lymph glands, liver, or spleen nor any palpable abdominal mass were found. Laboratory findings were: hemoglobin, 7.5 g/100 ml; reticulocytes, 1%; PCV, 27%; MCV, 76μ^3. MCH, 21, MCHC, 28%; platelet count, 740,000/cu mm WBC, 7500/cu mm, with a normal differential count. The erythrocytes were hypochromic and microcytic. Bone marrow biopsy showed abundant large particle iron stores mainly in the reticuloendothelial cells and no ring sideroblasts. Serum iron levels were 30 and 50 μg; TIBC, 291 and 222 mg/100 ml; fetal hemoglobin, 1%. Fecal urobilinogen level was increased. Autologous plasma was incubated in vitro with 20 μC of ferrous citrate ^{59}Fe of specific activity 9.7 μC/μg. Localization of organ reactivity was determined by surface counting. It was normal but there was a slow release from sacral marrow. Plasma and red cell volumes were determined by chromium labeling. Search for an occult tumor was started. Alkaline phosphatase level was BU 23 (upper normal 12 BU), and there was slight BSP retention. Electrophoresis showed albumin, 3 g/100 ml; alpha 2 globulin, 1.7 g/100 ml (upper normal 1.9/100 ml); haptoglobin, 220% of normal. Intravenous pyelogram and renal arteriogram showed a renal tumor. Liver scan showed slight hepatomegaly without filling defects, no metastases. A large noninvasive renal carcinoma was found and removed surgically together with the spleen and left kidney. There were no intraabdominal metastases. The patient received three units of blood; postoperative course was uneventful. Two months later all laboratory findings including ferrokinetic data had returned to normal: hemoglobin, 12.4 g/100 m, serum iron 113 μg; TIBC, 300, platelet count 386,000/cu mm; alpha 2 globulin, 0.7 g/100 ml; alkaline phosphatase, 12 BU; BSP, 2%. The patient has remained well for 2 years. Such studies are quite rare and should be repeated.

There are two other important types of anemia—iron deficiency and hemolytic. Simple iron deficiency anemia is connected with chronic bleeding. It will not be discussed here, but it is important to note that administration of large doses of iron may cure or at least improve the sideropenic anemia in such patients even though a cancer is present. This fact is therefore no argument against the presence of cancer. Iron deficiency anemia in elderly men is always suspect of gastric carcinoma and is quite rare without chronic bleeding (hemorrhoids).

Hemolytic anemia is much more complicated and has been the subject of much study. As a truly paraneoplastic condition, it is so rare, however, that very few physicians even with a wide experience will be able to get any extensive personal material. (I shall not discuss instances of hypersplenism in the lymphomas, where splenectomy—chemical, radiologic, or preferably surgical—may cure the anemia. In a way this is also a paraneoplastic condition.) Hemolytic anemia may occur in some carcinoma patients, and the reason for this is completely obscure (247). A number of cases of severe *hemolytic anemia and ovarian tumor* have been observed.

The most convincing observation of a true connection between acute hemolytic anemia and carcinoma was published by Barry and Crosby in 1957 (23).

The patient, a 26-year-old woman, had a very severe hemolytic anemia and showed numerous normoblasts in the peripheral blood. The Coomb's test was positive and "warm" agglutinins were found. No serum electrophoresis was performed. A tumor measuring 10 × 15 cm was removed from the ovary. Histologically, it was a teratoma and did not contain any visible lymphatic tissue or bone marrow cells. The enlarged spleen diminished considerably, and the anemia was cured but the Coomb's test remained positive.

Other patients with ovarian teratomas and hemolytic anemia also have been cured by surgery. One patient had a pseudomucinous cyst in an adenocarcinoma and also dermatomyositis. In several cases cited by de Bruyère et al (69) early splenectomy did not produce any lasting improvement, whereas extirpation of the ovarian tumors gave a complete remission. Treatment with glucocorticoids was usually ineffective. Most of the carcinomas were found in patients who had teratomas, but also cystadenocarcinomas have been described! (16) Autoantibodies, if present, usually disappeared after surgery. It is difficult to understand the mechanism of this effect even though some authors believe that the lymphatic tissue in a teratoma may produce immunoglobulins. The complication must be rare in ovarian teratoma since this is not an uncommon disease. In one of de Bruyère's cases the dermoid cyst contained antibodies with Rh specificity in a concentration 10 times that found in the serum. Therefore, it seems probable, that the tumor tissue may produce antibodies with hemolytic activity.

Refractory Anemia with Tumors of the Thymus

The connection between blood disorders and thymoma has been studied extensively. Most interesting from a therapeutic point of view is the fact that so-called pure red cell anemia, a variant of aplastic anemia, may be cured after the removal of the thymic tumor. It seems probable that the anemia should be regarded as paraneoplastic. It is probable that the first observation of the connection between severe anemia and thymoma was published in 1928. In 1939 Opsahl observed a man with hypoplasia of the bone marrow and pancytopenia along with a thymoma (321). In 1945 Humphreys and Southworth observed a 58-year-old patient who had had a mediastinal tumor treated by irradiation for 3 years (208). Later, pure red cell anemia developed. Thymectomy revealed a thymic tumor with a predominance of spindle-shaped cells. There was complete remission postoperatively. Since that time a number of excellent remissions have been observed, but it is also remarkable that several cases have had no or incomplete remission after thymectomy, whereas later splenectomy and treatment with ACTH cured the condition. In some well documented cases there was no effect at all after thymectomy. It seems as if we have no way to determine the type of patient who will respond to thymectomy (21).

A review of 56 cases from the literature including three personal observations was published in 1967 by Hirst and Robertson from Sidney, Australia (199).

Havard from Saint Bartholomew's Hospital in London published an excellent review of this problem (186, 187). During the last 20 years his group has seen 47 patients with myasthenia gravis and thymic tumor, but only four patients with this tumor had pure red cell anemia. In other hospitals with large patient populations, the latter condition is also very rare. It seems as if a certain anatomical structure could be connected with different symptoms. Castleman feels that myasthenia is rare in patients with spindle-cell thymomas, whereas ectopic production of ACTH is only seen with epithelial thymomas. Refractory anemia may occur together with different histological patterns, but it is believed that the spindle-cell type is most common, usually with added lymphocyte infiltrates. Hematologically reticulocytes are absent or very much reduced. Several patients initially had a pure red cell anemia and later developed leukopenia and/or trombocytopenia (128).

The fact that androgens may be effective in many patients with aplastic anemia is well known. Some patients with thymoma have benefited from this form of treatment, but this fact is not necessarily a proof that the anemia is not connected with the thymoma. For the present, it must be said that the mechanisms responsible for severe anemia in a few patients with thymoma are completely obscure. It is necessary, however, to look for thymoma in all such patients. In some hospitals the technique of visualizing the thymus by insufflation of oxygen into the mediastinum has given excellent pictures. The technique however, is usually regarded as difficult.

Hemoglobin F

The following case history may be of interest from the paraneoplastic point of view if future cases are discovered with the same constellation of symptoms (317).

The patient, a man born in 1918, had been treated for some years for refractory anemia, first detected in 1962. When we saw him in February 1967, very severely macrocytic with marked aniso- and poikilocytosis. He also had neutropenia but no thrombocytopenia. His serum iron level was high normal and he had a high concentration of carbon monoxide (CO) hemoglobin, indicative of increased hemolysis. Bone marrow biopsy showed hyperactive erythropoiesis but no ring sideroblasts. Hemoglobin analyses showed abnormal levels of hemoglobins F (35%) and A_2 (0.36%); no other pathologic hemoglobins could be demonstrated. No form of treatment was effective. His WBC count decreased, and the differential cell count showed low normoblasts. In January 1969 he became much worse, his hemoglobin level was only 4 g/100 ml. When he died, he had a large tumor in the right lung with liver metastases. Microscopically this was a bronchogenic squamous cell carcinoma and metastases were found in the intrathoracic lymph glands, liver, and spine. Erythropoiesis was not noted in the liver or spleen. Professor Betke, from Munich, studied the distribution of hemoglobin F in the red cells and found that it was heterogenous, that is, there were high, medium, or very low hemoglobin F concentrations in different cells.

There were no indications of foreign ancestry. When the patient's parents died, they had no anemia. His two sisters were alive, and their blood was examined electrophoretically. No increase in fetal hemoglobin, was found and the A_2 fraction was normal. The patient's four children were examined hematologically, and hemoglobin electrophoretic studies showed normal levels of hemaglobins F and A_2 in all. (Nyman, et al discuss the possibility that the patient's tumor may have activated the formation of fetal hemoglobin [317].)

In 1971, in Brisbane, Stewart published an account of a patient who had refractory anemia and a hemoglobin F level of 20% with hemoglobin $A_2$2,3%. The patient died and was found to have a malignant hepatoma but no signs of cirrhosis.

A group of investigators in Canada (Killen, et al) studied the incidence of increased hemoglobin F levels in a population of 700 cancer patients. The incidence in different groups varied between 2 and 27%, but the highest percentage was found in a group of 219 patients with ovarian carcinoma (32%). It must be regretted that the individual data are not given.

Another interesting study regarding disturbed synthesis of globin chains was published by Weatherall, et al, in 1977. This group studied a man of 81 who suffered from "a bizarre myeloproliferative disease which terminated in acute myeloblastic leukaemia." Ninety-five percent of his red cells generated typical hemoglobin H β_4) inclusions. His hemoglobin H was 50-60% of total hgb. There was no major deletion of the α chain genes in the DNA. "One possible explanation is that there are specific repressors of the α chain loci which are normally active only in embryonic life before α chain synthesis commences, and that these repressors had been activated in the leukaemic cell line."

Therefore, this could be another asynchronous depression resulting in a fetal pattern.

B. Erythrocytosis

At first it seems somewhat paradoxical that we should know so little about the mechanisms of the very common anemia in cancer, whereas we seem to know so much about the exact opposite, the much rarer erythrocytosis seen in some tumor patients. After all this is not so surprising since the substance stimulating erythropoiesis has been studied intensively for several years. The differential diagnosis between erythrocytosis and polycythemia vera (p.v.) may sometimes be difficult. An important diagnostic criterion of p.v. is splenomegaly. It is also common to find leukocytosis, sometimes with the presence of an increased percentage of basophilic leukocytes. These findings are usually incompatible with tumor erythrocytosis, even though such patients sometimes may also have leukocytosis. Thrombocytosis is common in both conditions, therefore this symptom can not be used for or against the diagnosis of a paraneoplasia. The red cell values in tumor erythrocytosis may be quite high, and in such cases, the physician may forget the possibility of a primary tumor.

In 1946, Jarl Forssell published a report of a case with "polycythemia" that disappeared when a renal carcinoma was removed. Blood and bone marrow findings remained normal for 14 months, when a contralateral renal cancer was detected and a relapse of the polycythemia occurred. Since then many cases of polycythemia have been described. From 1946 to 1958 Forssell personally saw nine additional cases, five had renal carcinomas and four had cystic kidneys (144). Forssell is without doubt the first to make these observations but his work is rarely quoted.

Published histories of patients with "symptomatic" erythrocytosis are now quite numerous, and it is clear that single cases nowadays are regarded as more or less uninteresting from the scientific point of view. In a recent survey of 340 cases of erythrocytosis, 179 had a primary disease in the kidney; 64, in the liver; 50, in the CNS; 25, in the uterus. Isolated cases have been described with adrenal, ovarian, pulmonary, or thymic tumors. The two leading groups have special interest, i.e. kidney and c.n.s. The problem regarding liver tumours is complicated (266). It seems as if a number of such patients were reported because they had shown much better blood values, i.e. less anemia, than might be expected from the fact that they had carcinoma! This may of course be true but it is not a convincing proof of erythrocytosis. It is therefore possible that the liver should come as number 3 or lower on the list (96).

Among the 179 patients with renal disease 120 had hypernephromas; 35, cystic kidneys, 3, Wilms' tumors; and the rest, various diagnoses. Somewhat confusing is the fact that no less than 14 of the patients had hydronephrosis only. This of course has nothing to do with paraneoplasia, and this problem will be discussed later when the mechanism of this condition is treated.

In a few patients with renal cysts it is difficult to understand how erythrocytosis could be a true paraneoplastic phenomenon. An interesting case report was presented by Vertel et al from the Michael Reese Hospital (450).

> The patient, a man of 47, had a hemoglobin level of 18 g/100 ml and a hematocrit of 60%. X-ray films demonstrated a large mid-renal mass with distortion of the renal calyces on the surface of the kidney. Sixty milliliters of fluid were aspirated. The cyst was closed but the walls were not removed. Erythropoietin was present in the fluid and in plasma and urine before surgery but not after the operation. Six months later all the values were normal.

This seems to be a clear-cut example of an actual connection, but the mechanism may well be discussed.

An excellent monograph on paraneoplastic erythrocytosis was published by Thorling in 1972 (436). This author has studied the instances of truly paraneoplastic erythrocytosis in great detail and has found that not only hypernephromas but also some other tumors—for example, hemangiomas, sarcomas, and Wilms' tumors in the kidney—have been combined with erythrocytosis. These cases are discussed critically. Regarding the combination of a hepatoma and erythrocytosis, the author points to the fact that erythropoietin determinations are very difficult in these conditions. Thorling has collected data on 23 cases of erythrocytosis and leiomyoma of the uterus. In 21 of the 23 the tumor was very large. All 23 patients had undergone surgery and in 18 of the 20 patients who survived a complete remission was seen. There was no relapse during the period of observation. Myoma of the uterus is extremely common (Some statistics indicate that it occurs in about 25% of the women.) A mere coincidence might therefore have been possible, if it had not been for the excellent results of therapy (204). In one of the two patients with no remission the spleen was enlarged, and this was obviously a case of polycythemia vera. It is difficult to draw any conclusions from the very few attempts at erythropoietin assay (504). The author discusses the possibility that these very large tumors may have a direct pressure influence on the kidney inducing increased erythropoietin production in this organ. The weight record is probably held by a patient published by Sjöstrand. The tumour weighed 11.5 kg and may well have had some mechanical influence on other organs.

Increased erythropoietin, (e.p.) production has been found in many cases of solid renal carcinomas. This is perhaps not so surprising as most authors believe that e.p. is produced by cells in the kidney. There has been much discussion about the different cells that may produce this substance, but the fact remains that kidney tissue contains e.p. and that anephric persons develop anemia and lack erythropoietin. In 31 of 33 cases with Wilms' tumors elevated e.p. levels returned to normal after removal of the tumor. The operative results were equally good in cerebellar tumors and in uterine fibroma, 26 of 26 and 24 of 25 respectively. Tissue extracts from many of these tumors gave positive results for e.p. Positive results were obtained in only two of nine hepatomas and in two of six uterine tumors. Most striking is the fact that ectopic production of e.p. exists only in the tumors I have mentioned.

Even though the hypothesis that the tumor tissue should produce e.p. is usually accepted (479) there are other possibilities that must be taken into account in the case of renal cysts and polycystic kidney and also in hydronephrosis (219). No neoplastic tissue is present that might be suspected as producing a special hormone. Several patients with such lesions have been cured of erythrocytosis after surgical treatment. Erslev believes that local anoxemia in the kidney from pressure on vessels may account for some of the erythrocytosis in these cases. I feel that many more critical and complete clinical observations are needed before this problem can be solved. I have seen a patient who had been operated on for a renal cancer. Her erythrocytosis disappeared but recurred, when she showed pulmonary metastases. Unfortunately there is no available evidence on the presence or absence of metastases in the remaining kidney. Local pressure in the second kidney may have been the explanation for the recurrence of erythrocytosis.

Regarding the frequency of tumor-induced polycythemia we have a few analyses from different hospitals. Our own experience at Malmö during a period of about 20 years has shown that the condition is rather rare. We have seen four patients with the combination of renal carcinoma and polyglobulia, but in only one patient did radical surgery give excellent results.

> The patient was a woman born in 1892. Blood values from 1954 were normal. In 1965 she had a high RBC count and a high hemoglobin level, whereas her WBC and platelet counts were normal. Her spleen was not palpable. A renal carcinoma was found and she was operated on in December 1965. She has been followed for 11 years and her blood values have remained normal.

In this patient the polyglobulia was the signal for the presence of a renal treatable tumor. In the three other patients, however, metastases were already present.

Berger and Sinkoff analyzed all the cases of hypernephroma from the Mount Sinai Hospital in New York. There were 273 cases, 2% of which had polyglobulia and 3%, amyloid. Two of the latter cases had started with fever. Another patient with an unexplained fever was mentioned. This patient died, and at autopsy hypernephroma with secondary amyloidosis was found (37).

In a study from Presbyterian Hospital in New York renal carcinoma was found in 4.4% of 205 cases with increased red cell counts (100).

The most enigmatic combination is that of erythrocytosis and cerebellar tumors (478). Fifty cases have been reported. The tumors were almost equally distributed between the two sexes, but erythrocytosis occurred in only five women out of the 50. The erythrocytosis was usually very pure without accompanying thrombocytosis. The spleen was never enlarged. Thirty-seven patients underwent surgery; in three a radical operation was impossible and another three died soon after the operation. Twenty-six have been observed for a long period, and complete remission of the erythrocytosis has been noted in all. Relapse of the erythrocytosis indicated recurrence of the tumor. Thorling (436) discusses the possibility that a tumor may produce ectopically the same hormone that is a topic product of the kidney. Much more critical work is necessary to understand these connections. Several authors believe that the cerebellar cyst fluid contains increased amounts of e.p. Some biochemical similarities between the erythropoietic factor prepared from the cerebellum and renal e.p. have been found.

The following case illustrates the fact that polyglobulia may be present with typical Hippel-Lindau's disease.

In one patient, a 12-year-old black boy with a RBC count of 8–8.5, a tumor in the median part of the cerebellum was only partially removed. The tumor was examined by electron microscopy. Different cell types were noted, but the pericytes were probably the most important. (Many authors regard these cells as endothelial in origin, and by electron microscopy, some of them have been found to contain what appear to be secretory granules.) This condition has been diagnosed as hemangiomablastoma. Kawamura et al state that the polyglobulia disappeared in half of the cases that were treated by surgery (228).

A priori we should have reason to assume that other carcinomas—for example, bronchial, mammary, and ovarian—should be capable of producing ectopic e.p. There are two possible explanations for the negative findings: (a) Erythropoetin cannot be produced as a truly ectopic product in these tumours. (b) Many tumors producing e.p. without erythrocytosis may have such a strong negative "anemic" influence on the patient's organism that this counteracts the effect of the erythropoietin. Determination of this substance in a large number of patients with carcinoma of the lung with normal or anemic blood values possibly may provide an answer to the problem (139). (In the connection it is interesting to note that a large number of patients have increased ACTH levels without clinical signs of Cushing's syndrome.)

Burgert et al found an *antierythropoietin* factor in the serum of an 11-year-old child with a mesenteric lymphoid hamartoma associated with refractory anemia, hypergammaglobulinemia, and retarded growth (71). The patient's bone age was 6 years. Seven days postoperatively the serum did not inhibit iron incorporation, and after 3 months the anemia was corrected, the gammaglobulin level had become normal, and the child had increased 2 cm in height.

C. Eosinophilia and Leukocytosis

Eosinophilia in tumor patients has long been known and discussed. Already in 1893 a German author published a report on a patient with a malignant tumor and a WBC of 120,000/cu mm 40% of which were eosinophils. Since then a large number of case histories has been published but very little information about the mechanism is known. Isaacson and Rapaport pointed out the fact that eosinophilia was not so rare (212). Of 15 tumor patients with eosinophilia, the majority had a tumor in the GI tract. The only patient with a bronchogenic carcinoma had a WBC of 80–140,000/cu mm with 18–32% eosinophils. One patient with a carcinoma of the colon with eosinophilic infiltrates had a WBC of 9,000/cu mm with 10% eosinophils. Five months after resection, his WBC was 8600/cu mm with 1% eosinophils.

The presence of a specific eosinophilopoietin has been proved by a very interesting recent study by Mahmoud et al in Cleveland (278). This group prepared a specific antimouse-eosinophil serum that did not work against neutrophils. They injected this serum into a group of mice and were able to deplete the eosinophils in their blood. Mahmoud et al followed the course of eosinophil regeneration after depletion and found that serum from these mice producing an excess of eosinophil cells contained a factor that stimulated eosinophilia when injected into untreated animals. They were also able to increase the incorporation of thymidine into bone marrow eosinophils. Attempts at purifying the poietin showed that it had a molecular weight of 186–1357 daltons on Sephadex filtration. It was dialyzable, heat labile and digested by pronase but not by trypsin.

It seems probable that this is a specific peptide that may well be produced ectopically by cancer cells and thus explain the marked eosinophilia in some cancer patients.

Another aspect of this problem was treated in a paper by Wasserman et al in (483) 1974 with the title "Tumour associated eosinophilo-tactic factor" (474).

These authors observed a patient, who was admitted with a WBC of 43,000/cu mm and 48% eosinophils. The highest value during the period of observation was 113,000 and 83%. The patient suffered from an inoperable tumor of the lung. At open biopsy 325 mg (wet weight) was obtained. It was extracted according to the technique used for the preparation of eosinophilo-tactic factor. When the patient died, larger amounts were extracted from the tumor, and the extract was purified on Sephadex GG 25; the molecular size of the extract corresponded to 500 daltons.

This substance was thought to be the same as the "normal" eosinophilotactic factor found in the lung. The tumor had a very low histamine content and no tissue mast cells were found. It was remarked, however, that there were rosettes of eosinophil cells around the cancer cells in the tumor. When tested in a micropore cell filled with the factor it was found that this cell attracted a large percentage of eosinophils from a mixed leukocyte population. Unfortunately the IgE was not determined. The pathologists felt that the patient did not suffer from eosinophilic leukemia.

The study is too recent and the technique probably too specialized to expect rapid confirmatory reports from other sources. The authors believe this factor should be compared with ectopic ACTH and PTH.

In the same journal two letters appeared on eosinophilia in tumor patients. In one letter, Dellon writes of two patients, one of whom had an eosinophil count of 784/cu mm before surgery, 164/cu mm after the operation, and later 1206/cu mm. In the second letter Healy from University College, Dublin gives data from a study of 140 patients with pulmonary carcinoma. Eleven percent of the patients had eosinophil counts over 500/cu mm. The histologic type of the tumor was not characteristic, but it was noted that no eosinophils were seen around the tumor cells. Also, no signs of allergy or drug reactions were present.

D. Thrombocytosis

The interest in platelet values in tumor patients has always been remarkably modest. In the literature there are only a few papers on platelets in carcinoma and even these studies are usually quite brief. Levin and Conley studied 82 patients with platelet counts over 400,000/cu mm who were found among approximately 14,000 patients from the wards of the Johns Hopkins

Hospital. Of these cases, the biggest single group (32) had a neoplasm, 18 had connective tissue disease, 4 had myeloproliferative disease, 5 had elevated counts as the result of acute bleeding with regeneration, and the elevated counts in the rest were due to various causes. Among the 32 tumor cases there was no patient with acute bleeding episodes. The tumor localizations were seven, gastric; six, colon; five, lung; two, breasts; two, ovary, and the rest from different organs. Only two patients had platelets over 1 million/cu mm—one with ovarian carcinoma and one with retinoblastoma. The authors point out the importance of the development of thrombosis, but they do not discuss the possibility of thrombocytosis as a paraneoplastic phenomenon.

Mayr et al have published an extensive study of thrombocytosis in patients with malignant tumors (289). Four hundred and eight patients were investigated and 17% had values over 400,000/cu mm. If we consider only the patients who had had no previous chemotherapy, 7 of 10 had hypernephromas; 5 of 16, colonic, 12 of 54, bronchial, and 10 of 80, mammary carcinomas; and 6 of 36, Hodgkin's disease. If patients with myeloproliferative syndromes are analyzed in the same way, it is found that only 11% have thrombocytosis. Two of the patients with renal carcinoma also had polyglobulia. Unfortunately the authors do not discuss the effect of therapy in the cases.

E. Thrombocytopenia

A very important and unusual relationship between a specific tumor and low platelet values was first described by Kasabach and Merritt in 1940 (225). They observed thrombocytopenia in a child with a giant hemangioma. Since their publication about 100 cases have been reported, some of them with very severe thrombocytopenia. It is believed that the platelets are trapped in the hemangioma and therefore rapid destruction should be the actual cause of the thrombocytopenia. Shim (393) has published an excellent study of all cases in the literature and points out that the mortality of the patients studied was 21%. Surgical excision is the therapy of choice, and excellent results have been seen after such operations. Preoperative biopsy seems to be important. Some patients have had multiple coagulation defects and also fibrinogen depletion. Therefore, coagulation tests must be run before surgery.

In 1972 Sundström et al published the report of a case of thrombocytopenia in a 90-year-old woman (428).

> She was bleeding and considerably anemic, with a platelet count of 10,000/cu mm. Bone marrow smears showed normal myelopoiesis and sections showed striking few megakaryocytes. The serum contained irregular isoantibodies against red cells (Coombs). Other autoantibodies were not studied. The patient died from intracranial hemorrhage. At autopsy a well-defined tumor, the size of a tangerine, was found outside the pericardium in the mediastinum. Histologically the tumor was classified as a lymphoepithelioma with spindle cell predominance in whorls. (This is the most typical finding when pure red cell aplasia is present.)

It is regrettable that the bone marrow was not studied at post mortem, but it is surprising if a patient with ITP should have a small number of megakaryocytes.

2. DISTURBANCES OF BLOOD COAGULATION IN CANCER

Our knowledge of disturbances of coagulation has increased remarkably during the last three decades. The number of factors necessary for correct blood clotting is very large. They number from I-XIII, and it has become rather difficult to get an overview of the different disturbances in their function except in highly specialized laboratories. The great majority of the different coagulation syndromes belongs to the inborn errors usually caused by *defective* formation—or more rarely by a total lack—of some active protein, a so-called coagulation factor. It is therefore easy to understand that carcinoma does not cause a very large number of coagulation defects since paraneoplasia is usually characterized by formation of an "extra" polypeptide.

By far the most important coagulation defect is the severe bleeding caused by a process that is interpreted very differently by different schools of coagulationists. One interpretation views hypercoagulability—disseminated intravascular coagulation (DIC)—as the basic disturbance, and the other maintains that lysis of fibrinogen and fibrin is the explantion (222). As yet, there is no final answer. The situation is very complex, but it is now generally believed that intravascular coagulation may be the first stage that leads to secondary, "reactive," fibrinolysis. In some instances clots are found in many small vessels. This is then regarded as a proof of intravascular clotting and rightly so. In others there are no clots, only bleedings. The defenders of the intravascular clotting theory maintain that clots were present initially and then were dissolved by secondary fibrinolysis! More plausible perhaps is the assumption that in these instances primary fibrinolysis had occurred (430).

We may well ask why it is impossible to decide these matters by the effect of therapy. Intravascular clotting must be treated with anticoagulants, preferably heparin, whereas the opposite type of therapy, namely, antifibrinolytic agents such as epsilon-amino-isocaproic acid (EACA), should be used if fibrinolysis is the primary process (315). In reality both methods have produced excellent results. Consequently, it seems that the severe bleeding must be caused by a spectrum of conditions, where lysis is dominant in some patients and clotting in others.

To me it seems clear that the only way to solve these problems definitely is to collect a number of patients who are well investigated with all available methods of coagulation study divide them into two treatment groups, and compare the heparin-treated group with the one treated with antifibrinolytic agents (293). We have seen patients, who responded well to EACA, but there are a number of patients who have also responded to heparin. The clinician is in a very difficult position if he chooses the wrong treatment. It is easy to kill the patients, and many doctors including ourselves have been somewhat reluctant to treat severe widespread life-threatening bleedings with heparin.

At the first Florence conference on hemostasis and thrombosis in May 1977 two papers were presented that may contain the practical solution to this problem. Gaffney has investigated the split-products resulting from the digestion of cross-linked fibrin and of fibrinogen. It seems possible to distinguish between these two patterns. Edgington and Plow have studied the antigens that result from in vivo fibrinolysis and fibrinogenolysis. They have found that plasma can be directly and specifically analyzed for these antigens by radioimmunoassay. These authors have studied the effect of proteases present in the leukocytes, and regard this as one pathway of lysis. If these methods are easy to handle, it will of course become possible to decide if a given patient has intravascular clotting with secondary fibrinolysis or if he has primary lysis of fibrinogen. This will be the final solution to the problem of DIC.

When judging the different diseases that cause life-threatening bleeding in otherwise healthy persons intravascular clotting in the pregnant woman seems to be the most important. However, in oncologic conditions, the most common bleedings occur in a very serious disease, namely, acute promyelocytic leukemia.

Jean Bernard and his group were among the first to describe acute promyelocytic leukemia as a special clinical entity (39). One of the most striking symptoms they observed was a tendency toward fibrinolysis. This has been confirmed in all hematologic centers, and we have seen a number of such patients in our wards. Goodnight has collected data on 76 patients from the literature with this condition and tabulated the important coagulation findings (160). It is clear from his presentation that skin manifestations were noted in all the cases where detailed information about the site of bleeding was given. It is unfortunate, however, that no distinction was made between ecchymoses, indicating surface bleeding characteristics of clotting disturbance and purpura, which is characteristic of thrombocytopenic disturbances. Only one patient had a comparatively normal value for platelets (134,000/cu mm), whereas the majority also had very low platelet values. This is usually explained by the so-called consumption of platelets. Almost half of the patients had bleedings in the central nervous system (CNS). Thrombi obviously were not especially looked for in the studies cited. Our own experience is chiefly with acute leukemia and prostatic carcinoma. In my experience I have found it very hard to judge the effects of therapy in acute leukemia because of the fulminant course of the malignant process. Even in the most severe types of acute leukemia, the results of therapy have improved markedly, and a good first remission is no longer exceptional. To my knowledge there is no clear-cut information on the development of bleedings after different forms of treatment.

Another neoplastic condition that often causes the same complication is carcinoma of the prostate with metastases. Goodnight has also collected data on 24 patients with this disease. He found that the skin is the usual site of bleeding, but that bleedings are sometimes indicated by hematuria and hematemesis. The findings are still difficult to judge, and it seems probable that the pattern is variable even though patients with marked thrombocytopenia were rare and most had normal or near- normal platelet values (355).

The situation in prostatic carcinoma is different. Forty percent of the patients with carcinoma of the prostate have marked in vitro increase in fibrinolysis, but as always, it is difficult to decide if this lysis is primary or induced by intravascular coagulation. It does not seem impossible that the carcinoma cells produce a fibrinolytic substance that is secreted into the blood (36).

In other carcinomas bleedings are much rarer, and we have also in the coagulation laboratory in Malmö General Hospital seen only a few such patients (478).

Goodnight has collected 21 published cases with different carcinomas (six, gastric; three, pancreatic; two, lung; and the others mixed) and bleedings. The fibrinogen values vary considerably from patient to patient, and thrombocytopenia is noted as occurring only in a few patients. It is therefore impossible to draw a clear clinical picture from his data (160).

4
Vascular System

1. TUMOR-RELATED DISORDERS OF THE HEART AND VESSELS
A. Thrombosis

Thrombosis in cancer is long-known among classical paraneoplastic syndromes. It was first described in 1877 by Trousseau, who himself suffered from gastric carcinoma and recurrent phlebitis, and this is therefore really worth the designation Trousseau's syndrome. Isolated observations were published during the following decades. (For a bibliography the reader is referred to Fisher and Hochberg [140].)

It was thought that pancreatic carcinoma was second among the predisposing factors. Sproul in a study at Presbyterian Hospital in New York found that 31% of the patients with pancreatic carcinoma had multiple venous thromboses (257). However, several other authors are inclined to regard carcinoma of the pancreas as not being especially predisposing to thrombosis.

In a published report of 166 cases of peripheral phlebitis it was said that thrombosis may be the first clinical manifestation of carcinoma especially in the lung or pancreas. Fisher and Hochberg reported four cases of primary pulmonary carcinoma with thrombophlebitis that did not respond to anticoagulant therapy. In three of the patients the first diagnosis had been thromboangeitis obliterans. The fourth patient, a man of 66, had an attack of severe thrombophlebitis in the leg, and later carcinoma of the lung was found and extirpated. During a 9-month observation period he developed no new thrombophlebitis.

In 1961 Irving Wright et al (257) published a review of 1400 patients who were admitted to New York Hospital with the diagnosis of venous thrombosis that was not postoperative. In their study they found 19 patients with carcinoma of the lung and 10 patients with pancreatic carcinoma. Thrombophlebitis was found in 31 of 51 patients before the discovery of carcinoma. The resistance to anticoagulant therapy was striking. Phlebitis may persist or occur over long periods of time before the primary disase is found. The authors point out that migratory or recurrent development is common, that veins of the arm and neck may be involved, and finally that anticoagulants are ineffective. All these characteristics are atypical for common thrombophlebitis. It was clear that the thrombophlebitic syndrome may appear a considerable time before the detection of the tumor, and it may be useful as an early warning.

A number of authors are convinced that thrombosis and carcinoma are related. Whether thrombosis is caused by increased coagulability or by other factors has not been settled. Several authors have assumed that a large number of patients who suffer from mucus-producing tumors have increased coagulability. This should certainly be investigated more closely with recent laboratory methods. One group found that one-third of the patients who underwent surgery for abdominal or thoracic carcinoma developed evidence of thrombosis as opposed to only 10% in the remaining non-operated group. Kakkar et al have discussed the possibility that in a hospital population there is a special high-risk group for thrombosis (223). It is quite clear that clinical signs are inadequate in the exact diagnosis of this condition. Therefore, over 200 patients, who were to undergo surgery, were investigated with the radioactive fibrinogen test. Results showed that patients who had had a previous deep vein thrombosis were at the highest risk, and the history of previous embolism increased the risk still more. Those with varicose veins, as well as with malignancy, and obesity (in all age groups), were also at risk.

The number of patients who have been observed for recurrent thrombosis and thrombophlebitis in unusual localizations and then lost their thrombotic diathesis after operation is not large (296). Womack and Castellano have published a report of one such case (501).

> A woman, 43-years-old, had an attack of phlebitis in the right leg in January 1949. In May she had phlebitis of the right lower forearm, and in July she had phlebitis in the left calf. The diagnosis of ovarian carcinoma was then made, and at operation a papillary cystadenoacanthoma of the right ovary and a Brenner tumor of the left ovary were removed. She was followed for 15 months after the operation and felt well with no recurrences of her phlebitis.

Woolling and Shick, at Mayo Clinic, published a report on a patient (case 6) with carcinoma of the ovary who had had thrombosis of both legs. The patient is living and free from thrombophlebitis 3.5 years after surgery (502).

Perhaps an important finding is that thrombocytosis is not too rare in patients with carcinoma, and it is possible that in some cases it may be a factor that favors the development of thrombosis.

B. Nonbacterial Endocarditis

One of the most enigmatic among the paraneoplastic conditions is nonbacterial endocarditis (NBTE). (Another name for the disease is marantic endocarditis—an obvious misnomer since a number of patients with it are not cachectic [268].) This type of endocarditis has been known for a long time to be a complication of neoplastic disease, even though the mechanism is obscure and the frequency was not known until Rosen and Armstrong at the Memorial Hospital in New York made an extensive study of the syndrome (364). The name implies that the endocarditis is originally nonbacterial even though there may be secondary growth of bacteria on the already changed valves. It is also clear that NBTE tends to occur on valves that have already been changed by endocarditis. The letter T stands for thrombotic since the true lesions on the valves are regarded as being of thrombotic origin and consist of fibrin and platelets. There is no sign of

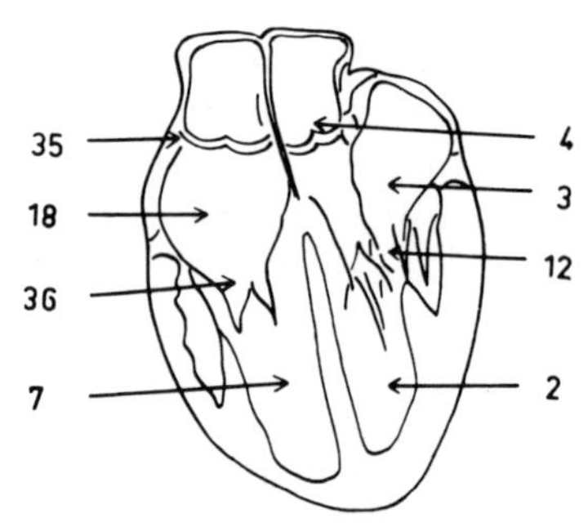

Figure 2. Schematic drawing of the heart giving number of patients with changes in different valves as described from the literature and from the author's own experience. The arrows pointing to the inside of the atria and ventricles indicate endocardial thickening. It is seen that a few patients also have changes on the left side. (From Thorson [437] 1958 with permission of the author and of *Acta Med. Scand.*)

inflammation and bacteria should not be demonstrated (except on the surface of an old lesion).

Rosen and Armstrong collected and analyzed data from 7,840 autopsies performed during 16 years at Memorial Hospital. Seventy-five cases of valvular endocarditis were found and discussed. Two-thirds of these had adenocarcinoma as their primary disease, the most common site being the lung and the least common, the breast. Hodgkin's disease and epidermoid carcinoma of the cervix and vagina were the only nonadenocarcinomatous neoplasias found along with the endocarditis.

Since NBTE probably may be regarded as an instance of diffuse intravascular clotting (DIC), it is interesting to note that only 10 patients had thrombocytopenia. Most of these had been treated with toxic drugs or radiation. In four patients abrupt development of thrombocytopenia together with thrombotic or embolic episodes was noted. It is therefore possible that intravascular clotting may be one cause even though the full-blown picture of DIC does not seem to have been observed.

Clinically, endocarditis is important because of the possible embolization occurring with it. Emboli were found in the spleen, kidney, brain, and heart, but it is probable that small emboli to the lungs were overlooked. The mitral valve was the primary site in 60% of the cases. In another 30%, a second valve was also affected. The pulmonary valves were only affected in one patient as was also the case with the tricuspid. Peripheral thrombophlebitis was never noted as the initial symptom even though some patients developed it later. It does not seem that cachexia was a prerequisite for the development of NBTE. Strangely enough there are no data regarding the frequency of non-neoplastic disease.

Rosen and Armstrong also analyzed clinical data from 30,000 general service cases at Memorial Hospital. The incidence of NBTE in non-cancer patients was 0.2–0.4%, whereas the incidence among cancer patients was practically 1%. It seems probable that this difference is statistically significant, even though it cannot be calculated from the available data. However, the data show that only ½-⅔ of all the patients had cancer.

C. Cardiac Myxoma

This is a very rare but important disease because it may be treated successfully by surgery. It is remarkable that these tumors often are accompanied by general symptoms such as weight loss, fever, and anemia (336). Patients with this condition often have a high ESR with an increase in gamma globulins. Studies have shown that they have polyclonal IgG. Another symptom that has been described is periods of numbness and pain in the fingers which may become white and stiff on exposure to cold. These painful periods may last for a few days and are separated by pain-free intervals. Cyanosis may also be present. We have seen one patient, who had this very markedly. When walking in cold weather, she suddenly developed complete right-sided hemiplegia and died of an embolism.

The most specific changes in the heart of paraneoplastic origin occur in patients with carcinoid tumors (see chapter 10).

2. TUMORS AND HYPERTENSION

A connection between renal tumors and hypertension is well known, and an incidence of 20 to 40% has been reported although it seems much too high. Severe hypertension, however, is rare, but it is now established that even severely hypertensive patients may be successfully treated surgically. It has sometimes been assumed that increased intracapsular tension might account for the high blood pressure, and this may be true in some children with nephroblastoma.

In 1949 it was stated that patients with nephroblastoma may have hypertension that should be cured after surgical removal or irradiation of the tumor that may recur when metastases appeared. Chisholm published reports on such cases in 1969 and 1971. In 1967 Robertson et al had found a patient who obviously had a tumor that produced increased amounts of angiotensin with secondary aldosteronism and hypokalemia (357). In 1964 six cases of hypernephroma with arteriovenous intrarenal fistula were described by Maldonado et al (279).

In these patients the diastolic hypertension was quite high and was abolished by nephrectomy.

> One patient had a renal carcinoma in the left kidney. Both the renal vein and the artery were dilated. When the artery was occluded the thrill disappeared and the vein collapsed. The kidney was removed. There were no signs of metastases. The blood pressure was only 140/70 mm Hg before the operation, but the heart was enlarged and the patient had left ventricular hypertrophy with an increased cardiac output. The results of the operation are not clearly stated.

Renal cell carcinomas that are quite vascular and contain arteriovenous shunts with a high cardiac output have been reported by Chisholm (81, 82).

A new complete syndrome was described by Conn et al in 1972 (85). These authors discussed all previously reported cases with juxtaglomerular cell tumors and added a new one.

> The patient was a young man with severe hypertension. He had hypokalemic aldosteronism and persistent hyperreninemia that responded to postural changes. A small renal cortical tumor was observed roentgenographically in the right kidney. Six days

> after nephrectomy the syndrome had disappeared. During the first 5 postoperative days there was marked tachycardia and systolic hypertension. Hypokalemia also occurred but was corrected with spironolactone. The tumor was regarded as of typical juxtaglomerular type. Electron microscopy showed numerous cytoplasmic granules in the typical tumor cells, and immunofluorescent studies with antibodies against renin showed beautiful fluorescence in 25-day-old cultures of cells.

Since the report by Conn et al a number of other case reports have been published (299, 380). Hypokalemia is a common complication and primary hyperaldosteronism was consequently expected. The hypertension was not always of malignant type. The tumor in itself is benign. Histologically, the neoplasm is considered to be a hemangiopericytoma and should arise from the juxtaglomerular cells.

It is possible that plasma renin levels in many children with nephroblastoma are higher than those in normal children, and that the level falls after nephrectomy. A child with malignant hypertension, high plasma renin level, and nephroblastoma was cured after nephrectomy (149). Thus, it seems that renin also may be produced by other renal tumors.

5
Kidney

1. MYELOMA-INDUCED RENAL DAMAGE

The classical example of a tumor that influences kidney function is plasmocytoma with production of Bence Jones protein. During the last decade we have learned that this protein, described in 1844 and for many years regarded as a curiosity, is one part of the immunoglobulin molecule. Studies of these so-called light chains present in the urine of patients with myeloma has taught us most of what we know about the structure of this part of the immunoglobulin molecule. The light chains are able to escape through the urine because of their low molecular weight. The complete immunoglobulin class G has a molecular weight of ± 150,000 daltons and consists of two identical pairs of polypeptide chains: the two heavy chains and the two light chains with a molecular weight of ± 25,000 daltons. Originally it was believed that these light chains were simply filtered through the glomerulus and then passed out with the urine through the tubule without causing any trouble other than the formation of precipitates within the lumen. This was the so-called theory of stopped pipes, and this mechanistic explanation of anuria or chronic renal damage in patients with myeloma was regarded as established. We still presume that this as an important factor, but it is possible that other mechanisms may be just as important. If there is no excretion of Bence Jones protein through the glomerulus, there will be no risk of developing myeloma-induced renal disturbances and uremia (406).

Some very rare cases of myeloma with a peculiar type of disturbed renal function have been described (127, 398). These patients have an acquired Fanconi syndrome, i.e., they have glucosuria, phosphaturia, and aminoaciduria. This indicates that the proximal renal tubules are damaged. In approximately 250 patients with myeloma, who we have followed personally in Malmö, we have seen only two patients with this disturbance. Excretion of Bence Jones protein occurs in 30% of all patients with early myeloma, and it is not clear why such a low percentage of them should develop this special damage to the tubules. A definite connection also has been found between myeloma-induced kidney disorders, Bence Jones proteinuria, and hypercalcemia.

Careful metabolic studies by a group at the NIH have definitely proved that Bence Jones protein is actively metabolized in the kidney. This metabolism must take place in the tubular cells after reabsorption, and it is possible that many more patients excrete light chains through the glomerulus than we find by urinalysis. Many authors believe that this retrograde passage of the low-molecular-weight protein through the tubular cells is responsible for most of the renal damage.

From the therapeutic point of view it is important to realize that certain factors are apt to provoke anuria in myeloma patients. Low intake of fluid, loss of water through perspiration, possibly acidification of the urine, and hypercalcemia are all risk factors. It has been maintained, and rightly so, that intravenous pyelography may cause an acute renal shutdown. I have seen this in several patients, and I have also seen patients who had a pyelography without any serious effects. Many authors are of the opinion that recent developments with new radiopaque substances have abolished the risks. This may be true but the fact that the test involves dehydrating the patient and compression of the ureters is certainly an added risk.

For this reason I believe intravenous pyelography should be avoided. This means that a correct diagnosis should be arrived at very quickly because the indications for pyelography are of course other tentative diagnoses and never myeloma. One thing is clear. The treatment of myeloma kidney damage should consist chiefly in keeping the diuresis as high as possible without causing water retention. Daily accurate weighing of the patient is therefore necessary. If the impairment of renal function is not too severe improvement may be seen in almost all cases. The occurrence of myeloma-induced renal disturbances is, however, a serious prognostic sign in the long run.

The urine should be examined not only with a dipstick (Albustix) but also with another method, such as the Heller test, which uses concentrated nitric acid, or sulfosalicylic acid, for detecting proteinuria since Bence Jones protein does not give positive reaction on strip tests (404). The combination of albuminuria and excretion of Bence Jones protein is not uncommon.

2. NEPHROSIS

Other examples of humoral influences from carcinoma products on kidney function are rare (90). However, there have been a number of published reports of cases of nephrosis in cancer patients, and it has been pointed out that cancer is about 10 times more common in patients with nephrosis than it is in a normal population. It also has been stressed that cytostatic treatment may improve both the neoplasia and the renal disease. This cannot be regarded as proof of a causal connection since there are many patients with the nephrotic syndrome and no tumor who respond well to cytostatic treatment. This is definitely an example of immunosuppression. Many years ago, before the advent of immunosuppression therapy, we saw a patient who had severe nephrosis and also showed signs that were interpreted as indicating Hodgkin's disease. The patient was treated with

Mustine with excellent effect—on the kidney. It was later found that our suspicions about Hodgkin's disease were incorrect. I mention these findings to stress the fact that effect of treatment with cytostatics cannot always be regarded as convincing evidence of a true paraneoplasia. The best proof is of course reversibility after extirpation of a tumor. Such cases have been described, one of which by Cantrell (73).

The patient was a 60-year-old man with nephrotic syndrome from whom large polypoid ulcerating gastric adenocarcinoma was removed. After the operation the patient recovered rapidly and 19 days after, he had no proteinuria. The patient was completely symptom-free 10 years later (73).

A similar case was reported by Revol et al (351).

The patient had a bronchogenic carcinoma with gynecomastia and osteoarthropathy as well as heavy proteinuria but not the full nephrotic syndrome. Two weeks after surgery all the paraneoplastic symptoms had regressed, but they recurred 6 months later when metastases appeared. Cyclophosphamide abolished the proteinuria a second time.

Several authors believe that nephrosis in a patient over 40 should warrant a search for a hidden neoplasm (251).

Another interesting case is the following published by Humphreys et al from the Mayo Clinic (207).

The patient was a girl of 18 with a nephrotic syndrome for about one year. During the last days before admission she had gained 4.5 kg in weight and developed severe proteinuria. Serum electrophoresis only showed a low serum albumin, 23 g/L. On palpation an abdominal tumor was found. Laparotomy disclosed a tumor in the jejunal mesentery. It was circumscribed and mobile and did not affect the renal vessels. Needle biopsy of both kidneys was performed, and the tumor was removed.

The renal biopsy was examined by light, immunofluorescent, and electron microscopy. The lesions were interpreted as minimal change in glomeruli. The most interesting findings were in the tumor. This weighed 200 grams and was diagnosed as angiofollicular lymph node hyperplasia (lymphoid hamartoma). The tumor contained well-developed lymphoid follicles with germinal centers and other irregular aggregates of lymphocytes. The perifollicular tissue contained many mature plasma cells, and these are obviously a conspicuous cellular element in this tumor.

At the follow-up 2½ months after the operation the patient was well and had returned to her normal life. Two and a half years later she remained well.

Such tumors are rare. They have also been called giant hemolymph nodes or benign giant lymphoma. They may be found on routine thoracic roentgenograms and may be completely asymptomatic. The plasma-cell type may present with refractory anemia, increased ESR, and hyperglobulinemia. These symptoms may disappear after excision of the tumor. An interesting example of the activity in such a tumor was published by Burgert (71) and has been discussed in Chapter 3.

3. AMYLOIDOSIS

Amyloidosis of the kidney is rare in all tumors even in myeloma. We do not know why only a limited number of patients with myeloma develop amyloid. The distribution is usually of the so-called primary type. This term should be avoided since amyloid in myeloma very definitely is a secondary phenomenon. So-called primary amyloid, also called atypical, is not localized mainly in the liver, spleen, and kidney, like classical amyloid after chronic infections and in chronic rheumatoid, steroid-treated, arthritis. However, it is not extremely rare to find renal amyloidosis in a patient with myeloma, even though it may be difficult to tell if this amyloid causes severe functional disturbance. Other tumors are rarely connected with deposition of amyloid substance, and it is only the paradoxical development of bilateral renal amyloidosis in patients with unilateral renal carcinoma that is of some importance (234, 331).

4. POLYURIA

Polyuria is mainly a result of damage to hypothalamic centers and/or the posterior pituitary. This may be a mechanical effect of tumors. It should not be forgotten, however, that a condition that may be called nephrogenic polyuria may be paraneoplastic. All conditions with marked hypercalcemia may lead to an ADH-resistant defect in concentration of the urine. It is possible that this may be explained by precipitation of calcium salts in the tubules, but it may also have more subtle explanations since it influences the function of cAMP in the kidney and this is obviously most important for the regulation of water metabolism.

5. LYSOZYMURIA

In 1966 Osserman and Lawlor reported on patients with monocytic and myelomonocytic leukemia, who excreted large quantities of lysozyme in the urine (325). It was also noted in this study that several of the patients had significant hypokalemia, and it was presumed that there might be a causal relationship between lysozyme damage to the renal tubules, hyperkaluria, and hypokalemia. This problem has been further investigated by Osserman and his group (304) and in a paper in 1970 they discussed the relationship of lysozyme to nephropathy (324).

Lysozyme has a molecular weight of 14,700 daltons and therefore passes easily through the glomerulus into the filtrate. It is probable that the enzyme is to a large extent reabsorbed in the proximal tubules. There is no question about the fact that there is a connection between hypokalemia and marked lysozymuria, but the mechanism is not quite clear (238). Klockars et al found that a special strain of chloroleukemic rats develops production of large amounts of lysozyme (238). This accumulates in the kidney and is excreted in the urine. The proximal tubular cells contain large amounts of lysozyme with characteristic droplets seen in the electron microscope. If rat lysozyme is injected into the aorta of normal rats, the same structural abnormalities are not seen, and it is possible that the mechanism is rather complex.

In 1972 Skarin et al published the results of lysozyme determinations in a large number of patients with leukemia and related disorders (401). Three patients with AML showed a very marked increase in lysozyme excretion during periods of severe relapse. At the same time significant unexplained persistent hypokalemia developed, and in one case the patient died without correction of the hypokalemia. The authors stressed the connection between these symptoms but did not find any renal tubular changes at the postmortem examination.

Pruzanski and Platt studied the excretion of several proteins of different molecular weights in patients with monocytic and myelomonocytic leukemia (344). Nine had normal renal function and nine had azotemia. Lysozyme excretion was always high but was higher in the latter group. High lysozyme concentration in the glomerular filtrate may induce tubular damage. Combined glomerulotubular dysfunction is unique to these diseases. Serum potassium levels were repeatedly low in three patients with azotemia and in four with normal renal function. Three patients also had low serum calcium levels, and two of these also showed a low magnesium level. Attempts to correct the hypopotassmia by oral doses of potassium were unsuccessful. None of the patients had tubular acidosis, glucosuria, or aminoaciduria.

6. METABOLIC ACIDOSIS

Metabolic acidosis may be seen in some tumor patients. The organic acid responsible is always lactic acid. The syndrome is almost exclusively found in cases of acute leukemia or Burkitt's tumor in relapse. It has been shown that the leukemic cells produce lactic acid in culture. Patients, who have had ascites with a heavy tumor cell load, had a high production locally. The disturbance is only seen in advanced stages, and a good remission abolishes this metabolic situation. On the other hand it is clear that it is not only a quantitative but also a qualitative problem. Only a small percentage of patients (including those with a large tumor load) show this symptom. It seems clear that it is a true paraneoplasia as only a minority of patients with leukemia or Burkitt's tumor have the metabolic abnormality. How it should be explained is still a mystery, but glycolysis has clearly reverted to an anoxic pathway. Block has hypothesized that the leukemic cells are not readily deformed in a normal way, and therefore, when they pass through the microcirculation are plugging of the vessels and secondary hypoxemia results (48). This explanation seems difficult to defend, and it is clear that a special metabolic defect of certain leukemic blasts is responsible even if we cannot define that cell morphologically. Therapy consists of large doses of sodium bicarbonate, but it is also important to avoid overdosage. The immediate prognosis is serious if the condition is left untreated.

There are other causes of acidosis in tumor patients that must be recognized to be properly treated. Examples are an acquired Fanconi syndrome as seen in some patients with myeloma and excretion of light chains and the rare type of tubular damage seen in a number of patients with hypergammaglobulinemia, especially the type that is connected with purpura hyperglobulinemia (302). In the latter condition there is formation of protein complexes that are detrimental to tubular renal function. The exact mechanism is not known. Immunoglobulin complexes have been observed in tumor patients, but it is not known if there is a causal connection between these complexes and the tumor. Purpura hyperglobulinemia is related to the so-called autoimmune diseases and the connections between neoplasia and this group of conditions are obscure but much debated.

The problem of renal tubular acidosis and potassium loss in patients with tumors in other organs will not be treated here. It is clear, however, that this may be an interesting clinical picture that is hard to analyze. Hypokalemia in itself has a damaging effect on renal tubules. This is seen in patients with adrenal tumors with hyperproduction of aldosterone and also in some patients with pancreatic adenoma and severe diarrhea as well as in those with villous adenoma of the rectum with copious amounts of liquid stools. The damage to the kidney is usually not severe.

7. METABOLIC ALKALOSIS

A report of a unique case has been published with persistent hypokalemia and a relatively high urinary potassium excretion secondary either to a renal tumor or to an adrenal adenoma or to both. We should not forget the very interesting hypokalemic nephropathy combined with metabolic alkalosis that may accompany villous adenoma of the rectum. The high volumes of fluid that are lost contain potassium and this leads to severe depletion. The disease is radically cured by operation.

The diluting effects that are connected with the presence of inappropriate amounts of ADH secreted by tumors may lead to severe consequences with marked hyponatremia. This syndrome does not seem to be very rare and several hundred cases of it are known. Half of all published cases were bronchogenic carcinomas but tumors in the duodenum, pancreas, colon, ovary, and laryngopharynx also have been reported. Hodgkin's disease also has been observed (150). It is interesting that treatment with intravenous vincristine or cyclophosphamide has produced transient hyponatremia. The number of patients in whom ADH-like activity has been found in the tumor is increasing and half a dozen cases have been carefully investigated. In some of these, radioimmunoassay of the blood has shown elevated levels; in others, bioassay of plasma has given positive results. At least two patients have been cured of the syndrome by removal of a bronchogenic carcinoma.

A. Villous Tumors of the Rectosigmoid Region

In 1954 pronounced electrolyte disturbances were reported in a patient with diarrhea for 10 years and hypopotassemia who recovered after removal of a villous tumor from the rectosigmoid junction (see also 491). In 1959 Roy and Ellis described a syndrome caused by a villous tumor in the rectosigmoid region (370). This patient had severe hyponatremia, hypochloremia, and hypokalemia. Since then a large number of such patients have been described (for an extensive review see 389).

The usual complaint is watery diarrhea, and medical interest has been centered on the low potassium values with the term "potassium secreting tumor" being used as a popular designation. A relatively unremarkable external factor may precipitate severe marked weakness, and the patient then becomes acutely and very alarmingly ill. An addisonian crisis may be suspected but the potassium level is *low not high* (491).

Characteristically the tumor is always located far down in the colon, most commonly in the rectum or in the borderline zone between the sigmoid colon and rectum. The tumor is villous and often large with a soft surface and numerous papillae covered by villi. It may be difficult to palpate because of its softness. Usually (in 90% of the cases) the tumor is benign. The volume of secreted mucus may be very large. It is remarkable that the level of potassium in the secreta is very much higher than in the plasma. The fact that very little reabsorption takes place in the lower colon and rectum accounts for the large losses of electrolytes and water. Several authors emphasize that a considerable quantity of sodium is also lost. However, hypokalemia produces severe symptoms in different organ systems, and therefore potassium is the most important ion clinically.

This syndrome is obviously a paraneoplastic condition which may be life-threatening. Correct diagnosis and operation will be effective in the large majority of cases who have not developed any metastases. The diagnosis may be very difficult. Many cases have been misdiagnosed and it is probable that many cases still are not being treated correctly.

8. RENAL TUMORS AND THE LIVER

In 1961 Stauffer published a brief communication about "nephrogenic hepatosplenomegaly" (412). The five patients he described had renal carcinoma, hepatosplenomegaly, and disturbances in liver function without any liver metastases. This problem has been studied by many physicians, and it has been found that the symptoms may be reversible after nephrectomy.

Scherstén et al (381) have shown that elevated levels of serum alkaline phosphatase and different transaminases may be found in patients with renal carcinoma without any signs of metastases. It is impossible to state if the alkaline phosphatase seen in these patients is always of the Regan isoenzyme type or if it is typically hepatic. The problem of isoenzymes in the phosphatase group is too complicated to be discussed.

We have seen an interesting patient, whose case history was published by Axelsson, Hägerstrand, and Zettervall (11).

The patient, a 34-year-old woman, suffered from joint pains and subfebrility. Her ESR was chronically elevated. Her condition was regarded as "rheumatic," and she was treated with steroids without any definite effect. When we saw her, she had a persistent tachycardia, an ESR of 140 mm/1 hr, a hypochromic anemia, and an elevated platelet count (530,000/cu mm). Protein electrophoresis showed a very high fibrinogen level (10 g/L) and signs of inflammatory reaction. Prothrombin (proconvertin) activity was low (50%). Serum alkaline phosphatase was 17 BU (upper normal: 8 BU). Liver biopsy stained by enzyme techniques to detect alkaline phosphatase activity showed widespread canalicular activity. (For color photographs see [10].) Biopsy from lymph glands, muscle, and temporal artery showed nothing pathological. Comprehensive X-ray investigation including intravenous pyelography was negative, but a renal arteriography detected a tumor in the lower part of the right kidney. Operation disclosed a renal adenocarcinoma with predominance of clear cells. On the fourth postoperative day the alkaline phosphatase was still high

(19 BU), but 4 months later her ESR was 11 mm/1 hr, platelet count, 202,000/cu mm; coagulation factors, normal; and serum alkaline phosphatase, 5 BU. In April 1976 she had no pains and felt well.

Dr Hägerstrand has found a number of patients, who have had malignant tumors and this pattern of canalicular phosphatase increase in the liver (298). This pattern has been reported in 71% of patients with malignant tumor not involving the liver. The findings are not specific, however, since many patients with rheumatoid arthritis also show the same pattern. Thrombocytosis also seems to be common.

6
Skeleton and Ossification

1. Osteomalacia and Tumors in Soft Tissues
2. Calcification of Tumors

1. OSTEOMALACIA AND TUMORS IN SOFT TISSUES

The number of patients with so-called adult onset, acquired vitamin D-resistant, hypophosphatemic osteomalacia is very small. In the familial disease with similar symptoms the disease process starts early and is obviously connected to a defect in the proximal renal tubules with deficient reabsorption of phosphates. Prader and Uehlinger (1959) were probably the first to observe a favorable effect of tumor resection on this disease process (342). In 1964 Dent and Friedman saw a "spontaneous" recovery from severe "osteomalacia" in a patient after a tumor had been removed from the femur. A report of a case of the same disease was published by Yoshikawa et al in 1964 (508).

The patient was a 54-year-old woman with acquired osteomalacia without glycosuria or aminoaciduria. She was treated with very large doses of vitamin D, and her skeletal symptoms improved. She also had an ulcerating tumor of the left knee that could not be removed because of massive bleeding. However, the tumor, diagnosed as a cavernous hemangioma, was removed at a later date.

In 1970 Salassa, Jowsey, and Arnaud published a paper in which they described two adult men, who had made spectacular recoveries from severe osteomalacia after removal of small tumors (377). The first patient was admitted in 1952, and at that time nobody dared to accept the fact that the dramatic recovery from the bone disease could be caused by the excision of a tumor. When they saw a second case with the same metabolic disturbance and the same type of sclerosing hemangioma, they realized that this must be an ectopic humoral syndrome. Their theory is that the hypothetic substance affects the proximal renal tubules causing what has been called "phosphate diabetes." (Their paper contains excellent illustrations of X-ray films and microscopic specimens.)

In 1972 Evans and Azzopardi described another case (131).

The patient, a woman, had a fractured femur and osteomalacia with renal glucosuria and increased excretion of glycine. She was treated with very large daily doses of calciferol and oral neutral phosphates. After some months she was considerably improved and even was able to walk outside. However, her serum phosphate level remained rather low. A lytic lesion was found in the region of the fracture and was curetted. During the next 2 months her serum phosphate level increased considerably and her condition remained static.

The most important recent publication comes from Olefsky et al from Stanford University (318).

The patient was a man of 40 who had developed symptoms of osteomalacia with hyperphosphaturia, glycosuria, and some aminoaciduria. He was given very large doses of vitamin D and phosphates and responded dramatically but later developed hypercalcemia. Vitamin D therefore was stopped and he had a relapse of his bone disease. With large doses of vitamin D his condition again became fairly good. At that time a tumor of the right lateral pharyngeal wall posterior to the last molar was removed. After the operation his serum calcium values rapidly rose to hypercalcemic levels, and his serum phosphorus level also increased. Vitamin D was stopped and his serum calcium level remained elevated, but his serum phosphate stayed at a low normal level. A parathyroid adenoma was removed but hypercalcemia persisted. The patient was doing quite well, and his glycosuria had disappeared.

The tumor was composed of sheets of fibroblasts with numerous blood vessels. Some parts resembled myxoma, others angioma. Clusters of large multinucleated cells resembling osteoblasts were in some areas associated with cartilage and bone formation. These tissues were also found without giant cells and bone was not always associated with cartilage. The histological pattern was difficult to classify.

The authors had an opportunity to examine sections from the two tumors described by Salassa et al and also a section from a tumor removed from a patient with a similar condition in Boston. All four tumors were very similar: all had bone formation. Two also had cartilage and three showed no continuity with bone. (Prader's and Uehlinger's case also had a tumor with giant cells.) The tumors were obviously a very unusual type of benign mesenchymoma.

Regarding the pathophysiology most authors express the opinion that the tumor must secrete a vitamin D antagonist. (The last case published was clinically different from the others.) This disease is obviously important from the practical point of view. All cases with adult vitamin D-resistant osteomalacia should be examined for the presence of tumors in the soft tissues. At surgery the tumor should be properly handled for the preparation of the active substance. This also might be of great theoretical importance.

2. CALCIFICATION OF TUMORS

Tumor metastases may become calcified, and it has been long known that patients with generalized mastocytosis may have extensive calcifications in the bones. Because metastases from medullary thyroid carcinoma are sometimes seen as calcified lesions in the liver, it was thought that such calcifications may be connected with local production of calcitonin. We therefore investigated the serum levels of calcitonin in a man who had very widespread mast cell infiltrations and also localized osteosclerosis in the skeleton. This patient had no increase in serum calcitonin concentration. We also observed another patient with calcifications of pulmonary metastases and also osteosclerotic foci in the bones who was observed by us and has been described. She had a primary gastric carcinoid. Unfortunately, no serum is left for calcitonin determination.

In a paper 1972 Morgan et al discussed liver calcifications in two cases of primary hepatoma (300).

> A man of 40 with classical liver cirrhosis and a woman of 21 both had finger and toe clubbing and signs of true osteoarthropathy with a typical radiological picture in the bones. In the man rosettelike areas of calcification in the liver were apparent on X-ray films. Postmortem examination showed a poorly differentiated largely necrotic multicentric liver-cell carcinoma with extensive calcifications. No bone formation was seen. No tumor was present in any other organ. The woman also died and at autopsy tumors were found in the liver, spleen, and lung. The liver was very big, (3,670 g). A large calcified area had to be cut with a saw and smaller calcified tumors were found in the rest of the organ. The histological picture was consistent with a liver-cell carcinoma, but there was no evidence of cirrhosis.

Hepatoblastoma in children may become calcified, and it is thought that bone formation may have something to do with embryonic tissue. This problem was treated by Kasai and Watanabe in a study of 70 cases with liver carcinoma. Some of the tumors were regarded as embryonic. A number of patients had osteoid tissue in their tumor. It is probable that some substance produced by the tumor induces bone formation since such inducers have been identified. The fact that the two patients presented by Morgan et al both had osteoarthropathy and such massive calcification could mean that the two disturbances are in some way related since they are both very rare and a combination must be extremely uncommon if fortuitous. Primary osteogenic carcinoma of the liver has been described (220).

Part 2

GENERAL SIGNALS AND SPECIFIC SUBSTANCES INDICATING NEOPLASIA

7
General Symptoms and Immunological Markers

1. Fever
2. Elevated Erythrocyte Sedimentation Rate (ESR)
3. Fetuin
4. Carcinoembryonic Antigen (CEA)
5. The "Regan Enzyme"

1. FEVER

It has been known for a very long time that fever may be one of the symptoms connected with renal tumors. Ninety years ago a German author by the name of Stetter published a paper on this problem and it was discussed in some detail by Israel already in 1911.

In an analysis of patients with fever of unknown origin, Böttiger established that 10% of them were later found to have extrapulmonary tuberculosis, another 10% had Hodgkin's disease or hypernephroma, and only 4% had other carcinomas. It is therefore clear that hypernephroma is an important cause of occult fever. Böttiger has seen only five patients (one man and four women) who did not become afebrile after surgical removal of the tumor. He has also investigated the prognostic importance of fever, and found that there was no difference in mean survival time for the patients with fever and those without. The group with fever that disappeared after surgery had about the same prognosis. Böttiger also found that the incidence of metastases was higher in the febrile group than in the afebrile group, but presence of metastases did not change the effect of surgery on their febrility (54).

We know very little about the mechanism that is responsible for the development of fever in tumor patients. Böttiger quotes an interesting case of a patient who had been treated for a long time with steroids for febris causa incerta without any effect. When a renal tumor was detected and removed, he became definitely afebrile after 2 days.

Studies by Westphal, Wood, and others show that fever is commonly caused by a pyrogen released from neutrophilic leukocytes. Lymphocytes do not produce

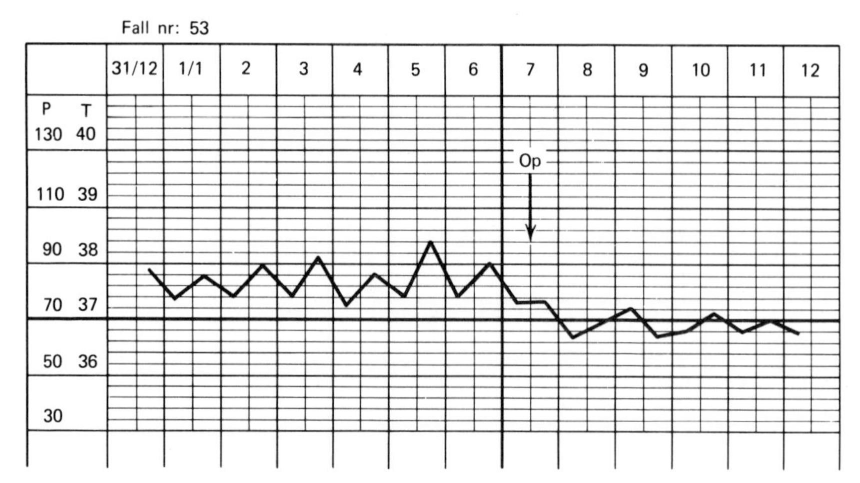

Figure 3. Effect of nephrectomy on body temperature of a febrile patient with renal tumor. (From Böttiger with permission of the author and of *Acta Med. Scand.*)

pyrogen. The sequence has been demonstrated to be the following: An antigen-antibody- complement complex may release the leukocyte pyrogen (just as bacterial endotoxin and some viruses do). This pyrogen is transported via the blood to the brain. The hypothalamus then exerts its influence on the striated muscles in such a way that increased heat is produced. Sometimes violent shivering or rigor is visible. At the same time vasoconstriction occurs in the skin and little heat is lost from the surface. How fever is mediated in patients with tumors is not yet clear.

Rawlins at St. Thomas Hospital in London has performed some experiments on the mechanism of fever in renal carcinoma (348). The clinical material consisted of two febrile patients and one who had no fever. Saline extracts from the tumors producing fever caused short rapid monophasic fever reactions in the rabbits. Extracts from normal kidney tissue as well as from the afebrile patient's tumor caused no such reaction. Attempts to demonstrate pyrogens in the leukocytes from the tumor patients were unsuccessful. He noted, however, that the tumor in the afebrile patient was most necrotic. (Some authors believe that necrotic products are pyrogenic.)

Similar experiments were performed by Bodel at Yale (51). She was able to produce fever with extracts from tumors but not from normal kidney tissue. Similar extractions have been performed with tissues from patients with Hodgkin's disease, but the results seem ambiguous. It is difficult to judge the importance of these experiments, but they should be repeated on a much larger scale.

Fever also may be a symptom in patients with *lymphomas and leukemias*. In the last condition fever is usually caused by infections in the last stages of the disease, when inflammatory changes in the upper respiratory tract are common. In the lymphomas it is often difficult to decide if febrile episodes are caused by liability to infections (insufficient production of leukocytes or immunoglobulins) or by the disease itself. Hodgkin's disease, however, is definitely one of the common causes of fever of unknown origin. The classical undulant type (Pel-Ebstein) is not common but is of great diagnostic value when it is present. Undulant fever in general has become rare since the eradication of bacillus abortus Bang.

In 1966 a study on the clinical significance of fever in Hodgkin's disease was presented by Lobell et al from Salt Lake City (265). The authors regard initial fever as almost always caused by the disease itself and not by infectious complications. Only one patient in the very large group studied displayed the pattern of Pel-Ebstein fever. The authors analyzed the prognostic importance of fever, but it did not give much information since most patients were studied at a time when active therapy had not yet been started.

In many publications it has been maintained that tumors of the liver should be a common cause of fever. However, the authors, usually do not speak about primary hepatoma but rather about cases of liver metastases from other primary sites. This makes the analysis much more difficult, for natural reasons, and fever from liver tumors is hardly ever reversible.

Browder et al have published a paper on the significance of fever in neoplastic disease (67). They have collected data on all cancer patients admitted to one hospital during one year and have tried to analyze the material. They found that 30% of these patients had no elevated temperature. In 38% fever was due to "infection or obstruction," it was felt that 16% had fever unrelated to the tumor and only 5.4% were thought to have "tumor fever." In this group the lymphomas dominated. The frequency of fever was found to be four times greater in patients with metastatic tumors.

The authors cite an interesting personal observation.

> A patient had metastases to the skin from a hepatoma and marked intermittent fever. Removal of the skin tumor was followed by disappearance of the fever. Cultures from the tumor and the blood did not reveal the presence of any microorganisms. Histologically the tumor was densely infiltrated with leukocytes (polymorphonuclear).

The authors discuss the fact that pyrogen may be produced by leukocytes. It seems tempting to assume that such a pyrogen may have been responsible for the reversible fever.

2. ELEVATED ERYTHROCYTE SEDIMENTATION RATE (ESR)

It is a common belief among many clinicians that patients with renal carcinoma should have a much elevated ESR. Already in 1946 Olofsson made a study of the ESR in this condition and found that most patients had a normal or only slightly elevated ESR. Böttiger published the results of his study of 190 cases in 1966. He found that 20% of the patients had an ESR under 15 mm/1 hr; 45%, 16–20 mm/1 hr; and 33%, over 20 mm/1 hr. From the practical point of view this means that an unexplained elevated ESR in an elderly person should always be suspected as indicating a renal tumor. Ljunggren et al studied the presence of the so-called C-reactive protein in patients with renal carcinoma and found that it was often increased (264).

Böttiger has presented some arguments that patients with clear-cell carcinoma should have more fever and a higher ESR than others. One of the great difficulties in correlating histology and the presence of certain clinical findings is the fact that so many tumors are mixed. In such instances it is of course impossible to divide the patients into distinct groups according to histological criteria.

3. FETUIN

During the last decade numerous publications on immunological methods for recognizing cancer have appeared. In 1964 Tatarinow reported the finding of an embryonal α-globulin in the serum of a patient with primary hepatocellular carcinoma. This protein was found to be similar to α-fetoprotein which had been found in the cord blood from human babies (38).

In 1945 Pedersen published the results of his investigations on fetal blood from calves (330). He was able to prove that the serum contained an α_1-globulin that he called fetuin. The electrophoretic mobility, sedimentation constants, and so forth of this new protein were determined. It is quite clear that this fetal protein in different animals belongs to the same class as the fetoprotein found in human hepatoma (see however 235). A very extensive investigation of the α-fetoprotein content in patients with various conditions has now been carried out. It was found that of a total of 5,301 patients tested, 380 had a malignant hepatoma and fetoprotein was found in the sera of 259 (68%) of them. Interestingly some gonadal tumors have been associated with high levels of α-fetoprotein. It seems that false-positive tests would be very rare but it also must be remembered that one-third of the patients with malignant hepatoma do not have the fetoprotein in their sera. The difference between hepatoma patients who produce α-fetoprotein and those who do not is not yet clear. It is not only a question of sensitivity since it has been shown in experimental animals that seropositivity is different in hepatomas caused by different carcinogens. Tests of 15 rats with aflatoxin-induced hepatoma were negative, whereas tests of 42 rats with hepatoma caused by another carcinogen were positive. The histological type of hepatoma in the rat does not seem to be important, and there is not a good connection between histology and fetuin production in human patients.

The clinical value of this test is very great not only in patients with liver cirrhosis, who are suspected of having hepatoma, but also in a suspected ovarian or testicular tumor, where a teratoblastoma with fetal components may be diagnosed.

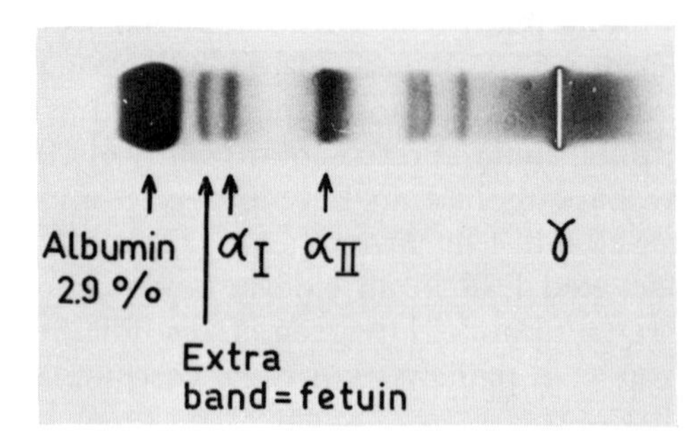

Figure 4. Electrophoretic strip of the serum in a patient with a malignant hepatoma. The α-fetoprotein (fetuin) is seen as a distinct band between albumin and α_1. The picture clearly demonstrates the migration velocity of the fetal protein. (From Trell et al. [444] with permission of the authors.)

4. CARCINOEMBRYONIC ANTIGEN (CEA)

In 1965, Gold and Friedman found that an antigen could be extracted from cancers of the colon that was not present in normal colonic tissue (156). Later it was found that other entodermal tumors contained this antigen and that it was wide-spread in the GI-tract. The most interesting part of the discovery was the demonstration of the same antigen in digestive organs from human fetuses at the age of 2–6 months. It is not present in the fetal lung or in bronchogenic carcinoma. The antigen has been named carcinoembryonic antigen (CEA). In a later paper the same group reported on the presence of CEA in the serum as demonstrated by radioimmunoassay. It is now possible to demonstrate presence of CEA in a large number of patients with colonic carcinoma, and it has become a widely accepted method for early detection of gastrointestinal cancer. Unfortunately, there are both false-positive and false-negative findings, but this may become an excellent procedure for the checking operative results and recurrences. (The interested reader should consult a recent review by Alexander [4] and also 371 and 416).

Theoretically this reversal to embryonal synthetic mechanisms is important. The authors in the nineteenth century discussed the parallel between fetal conditions and neoplasia. Their views were later criticized, but the old concepts now are more or less widely accepted.

5. THE "REGAN ENZYME"

In 1969 Stolbach and Fishman published the results of an investigation of an alkaline phosphatase isoenzyme that was present in the serum of 27 patients with various malignant tumors (141, 417). These investigators regard this finding as an instance of paraneoplastic ectopic polypeptide production. The isoenzyme was first found in a patient named Regan who had a carcinoma of the lung. Although the Regan enzyme has a number of characteristics in common with placental alkaline phosphatase, it is not considered to be identical with this isoenzyme. Only 13 of the 27 patients had elevated total alkaline phosphatase levels. The Regan enzyme is not inactivated by heating for 5 minutes at 65° C. Identification of the Regan isoenzyme in a specimen requires the finding of a placental band on starch electrophoresis as well as the absence of this band in aliquots preincubated with placental antiserum (312).

8
Specific Tumor-Produced Biochemical Disorders

1. INCREASED ACID PHOSPHATASE LEVELS IN PROSTATIC CARCINOMA

In 1938 the Gutmans already had established the fact that increased serum acid phosphatase values are common when there are metastases from a prostatic carcinoma. The primary tumor alone rarely produced a marked increase. These findings basically have been confirmed even though physicians maintain that the diagnosis usually is made on clinical grounds and not with biochemical methods. Despite the dissention, serum acid phosphatase determinations have been found to be an excellent procedure for following the development of the disease in many patients.

A critical survey of the problem introducing a more recent method for quantitation was published by Goldberg and Ellis (157). They arranged their patients into four groups based on Murphy's criteria for staging the disease. In group A were those patients in whom an occult unknown carcinoma was discovered at operation. Group B was made up of those patients whose tumor was confined to the prostatic gland. Those patients in whom there was local extension of the tumor to the paraprostatic tissue were put into group C. In group D were those patients in whom metastases to bone or soft tissue outside the pelvis were found. The skeleton of all patients was examined radiologically. In 60 cases the diagnosis was purely clinical. In another 52, histological diagnosis was established.

For each group the acid phosphatase levels were assayed using adenosine 3'-monophosphate as substrate. In group A, all patients had near-normal or at most borderline values; in group B (10 patients) only 2 had normal values; in group C (7 patients) all values were above normal; and in group D (21 patients) 2 had borderline values and all others had definitely pathological values. Included in the survey was also a large group of patients with benign hypertrophy of the prostate. In this group hypertrophy, with a few exceptions, the values were normal. It should be remarked, however, that in a group of 56 patients with other cancers, there were also 2 with high values.

On the whole, the results seem to be very reliable considering the fact that undetected or even undetectable prostatic carcinoma cannot be very rare. It seems reasonable to assume that the few patients in whom the values were high without positive diagnosis, in reality, had prostatic carcinoma.

It may be of interest to compare these results with those from other studies. All the statistics show that 80 to 90% of the patients with metastases, (group D) had pathological acid phosphatase values, and for those in group C, the percentage is about 60, if we accept only those values obtained by a reliable method. This seems to show the great importance of the method for the detection of prostatic cancer. There are, however, certain factors that have to be taken into account. One is that a false-positive test may result if a rectal examination is performed before taking the blood samples.

The reasons why serum acid phosphatase levels are elevated in prostatic carcinoma are not very clear. Carcinoma of the prostate contains *less* acid phosphatase than the normal gland, but metastases show a marked, increased activity as compared with the normal surrounding tissue, and the activity does *not* parallel the extent of metastases. It has been proved histologically that some neoplastic tissue may contain abundant enzymes, whereas the serum levels of the enzymes are normal.

Patients with skeletal diseases other than metastatic, prostatic carcinoma rarely show high serum acid phosphatase values with two exceptions. Both Paget's disease and hyperparathyroidism may increase the levels. In leukemia the acid phosphatase levels may be high because of the enzyme content of these cells. It is known that the lysosomes in many cells contain this enzyme. In a family in which there was an early neonatal death and one child had a complete lack of acid phosphatase in the lysosomal fraction of different cells, the trait has been described as autosomal recessive. These findings may be an indication that the enzyme is of vital importance. There are many isoenzymes of acid phosphatase. (It is possible that the prostatic enzyme is a different molecule.)

2. HYALURONIC ACID IN THE PLEURAL FLUID FROM PATIENTS WITH MESOTHELIOMA

In 1939 Karl Meyer et al demonstrated the presence of hyaluronic acid in pleural fluid from patients with malignant tumors involving pleura or peritoneum. Thompson et al (433) became interested in this problem and compared the content of acid mucopolysaccharides (AMP) in the body fluids from all his patients with malignant mesothelioma with the findings in samples of pleural fluids from 11 other patients. They found increased amounts of AMP only in the tumor patients. (One of their patients had a treated mesothelioma and showed negative findings.)

Rasmussen and Faber investigated pleural fluids from 247 patients. Seven of 19 patients with mesothelioma had a hyaluronate concentration of 1 mg/ml or more. A concentration higher than 0.8 mg/ml was found only in patients with mesothelioma, but it should be remembered that patients with bronchogenic carcinoma or with pulmonary tuberculosis also may have small amounts of this substance. For diagnosis, it may be important to run an electrophoresis of the fluid before and after treatment with hyaluronidase. The reaction may be negative in the presence of this tumor even though a number of investigators have maintained that it is diagnostic.

3. HYPERLIPASEMIA

Robertson and Eeles have discussed a patient with a syndrome that has been described in at least nine other cases (356, table). These patients were all men. They suffered from subcutaneous infiltrates that were somewhat tender and obviously were caused by panniculitis of the same type as Weber-Christian's disease. Arthralgia or arthritis were also common. Several patients had eosinophilia (maximum 35%). All these patients had had pancreatic tumors in different parts of the organ. Sometimes the primary tumor was quite large, and at postmortem examination numerous metastases in the liver were found. These contained increased amounts of pancreatic lipase, and a specific symptom was an increase in serum lipase. In one patient it was four times normal. No patient seems to have had the diagnosis intra vitam. It is therefore impossible to state if the condition may be reversible.

4. HYPOGLYCEMIA

The clinical symptoms of hypoglycemia may be rather uncharacteristic. Cerebral dysfunction occurring after maximal physical exercise or prolonged fasting is typical. The most characteristic symptoms occur in patients with insulin-secreting pancreatic tumors. In this disease there is an adenoma of proliferating beta cells from islet tissue. The insulin content of the tumor and blood is high, and all the symptoms disappear after removal of the tumor. This is a classical example of topic hormone production by a tumor. The disease is well known and will not be discussed here even though it sometimes may be quite difficult to recognize. (The reader is referred to textbooks on endocrinology [see also 309]. A number of excellent reviews treat the problem of Zollinger-Ellison syndrome [34, 164, 173].)

We find the same syndrome among a group of patients with different tumors, even when the localization is extrapancreatic. The mechanism has been much discussed but there has been no definite solution to the problem. One fact is obvious, however, and this is the large size of these tumors. Their localization and histological type is also rather characteristic. Among 180 tumors 40% were "mesenchymal" and either intraabdominal or intrathoracic. Twenty percent were different hepatomas (273) and the rest, in decreasing order of frequency, were (267) adrenal cortical cancers, gastric cancers, various lymphomas, and a number of tumors with different localization (397). Lung and ovary carcinomas were very rare. The localization of these cancers is thus quite different from most other carcinomas with paraneoplasia. An interesting suggestion has been advanced by Marks et al (283, 284). They found that the metastases of some definite beta-cell tumors were regarded by the pathologists as spindle-cell tumors. It is quite possible that some of the so-called spindle-cell tumors may have been such metastases.

Serum insulin has been measured in a number of these patients with very variable results. The same is true of the ILA content (insulinlike activity). A large number of metabolic studies have been performed to establish the response to glucose or tolbutamide or to determine the levels of growth hormone, free fatty acids, lactate and so forth. The results are inconsistent and disagreement about the mechanism is general. Surgical removal of large tumors sometimes improved or abolished the clinical condition. The ILA content may be lowered, but true immunoreactive insulin may remain normal in the same patient.

It is assumed that there is excessive consumption of glucose in the tumor. The experimental data are rather conflicting, but there are some reliable experiments that seem to indicate that certain tumors may have a very high turnover of glucose. Labeled glucose (^{14}C) had a very short biological half-life when given to a patient with a retroperitoneal sarcoma. In another patient with a mesothelioma, it was calculated that the tumor used several hundred grams of glucose in 24 hours (46). Kreisberg et al observed a patient with an adrenocortical carcinoma who required 400 grams of glucose per day to stay normoglycemic (243).

The patient, a 23-year-old woman, was admitted in coma with a blood sugar of 30 mg/100. Her glucose turnover rate was increased two-to threefold. She was markedly virilized, had hypokalemic alkalosis as a result of excessive androgen and cortisole production, and paradoxically, an excessive cortisone production. A left adrenal carcinoma weighing 800 grams was removed together with the kidney and spleen. Her biochemistry was restored to normal, and she became symptom-free and returned to work without having any special troubles. Six months after the operation she died in shock from tumor embolization of the pulmonary artery.

The authors interpret their findings as a combination of increased glucose utilization and decreased glucose production. They assume the presence of an unidentified substance acting on peripheral tissues and liver in a manner similar to insulin but undetectable by immuno- and bioassay technique. (Sixteen patients were known to have adrenocortical carcinoma and hypoglycemia at the time the case history was published).

Another patient, treated successfully, was a 47-year-old black man who had episodes of mental confusion and loss of weight, even though he was well nourished when admitted. Lowest blood sugar (24 mg/100 ml) was measured one day after he had missed his breakfast. A large retroperitoneal tumor weighing 790 grams and showing the histology of a spindle-cell fibrosarcoma was removed successfully. The patient was followed for 4 years. No symptoms were noted and he regained his lost weight (283).

It is very hard to express a definite opinion about the explanation of different causes for these disturbances of glucose metabolism, and this may account for the variable results that have come out of even very careful studies. Factors of importance are obviously consumption of glucose by the tumor, influence on the liver leading to decreased glucose mobilization or on the fat tissue with decreased lipolysis.

It should be pointed out that this probably is the only instance in oncology where it has been *proved* that the neoplastic cells cause symptoms by consuming necessary nutrients (so-called parasitism).

5. GLUCAGON-SECRETING TUMORS (HYPERGLYCEMIA)

During recent years the fact that glucose homeostasis is maintained as a balance between glucagon and insulin has been established. This has led to a much better understanding of many problems connected with glucose metabolism. A priori it seems probable that tumors secreting glucagon should be able to cause a syndrome that resembles diabetes. Such observations are quite rare, but it may be that the advent of radioimmunoassay for the determination of glucagon may be a help in diagnosing such conditions. Presently, we know that tumors with high glucagon production cause a clinical picture that is *not* dominated by the symptoms of hyperglycemia. In many patients this is even to be regarded as a minor detail in the clinical picture. Mallinson et al (280) have made an excellent survey of the clinical symptoms connected with hypersecretion of glucagon, and they regard symptoms from the skin as the most characteristic and therefore the most diagnostic (see also Chapter 1).

In 1966 McGavran, Unger, et al described a clinical picture that they explained by secretion of glucagon from a pancreatic tumor (274). In their paper they remarked about the peculiar changes noted in the skin and about the anemia. This case history discloses a completely new disease picture even if some parts were already known.

The small bowel was normal in this case, but it is known that glucagon inhibits motility in the human jejunum and colon (241). It is possible that part of the clinical picture seen in the patient may have been caused by a blind-loop type syndrome. Since that time a number of similar interesting observations have been published that cannot be quoted in detail. In a recent survey of the question, Mallinson, Bloom et al in London, discuss nine cases with the syndrome "necrolytic migratory erythema," stomatitis, weight loss, anemia, and usually some derangement of carbohydrate metabolism (280). All the patients had pancreatic tumors. Eight were undoubtedly islet-cell tumors; four contained glucagon and were probably derived from alpha cells. The metabolic effects in these patients are extremely interesting and probably were caused by the high plasma glucagon concentrations. The levels of amino acids in the plasma were very low, and it seems probable that this disturbance is responsible for the skin lesions and the anemia. Only one patient was diagnosed before death. In this case marked postoperative improvement was noted, and it is remarkable that the low levels of amino acids became normal or even somewhat increased after operation (see also 252).

For the first diagnosis the skin lesions are of course decisive. The rash may start on the lower abdomen, the groins, and perineum as an erythema with consecutive superficial blistering. After rupture these blisters leave crusts or a weeping surface. Healing occurs in the center with centrifugal spread. A gyrate pattern may develop. Hyperpigmentation is often secondary to healing. I have seen one patient personally who showed necrotic irregular areas on both legs.

Histologically the tissue shows necrosis in the upper layers of the epidermis. Later the picture resembles a nonspecific dermatitis. The differential diagnosis in these patients may be discussed from previous "labeling": benign pemphigus, pemphigus foliaceus, toxic epidermal necrolysis, subcorneal pustular dermatitis, generalized pustular psoriasis and so forth. Most patients were postmenopausal women. The diagnosis rests on the measurement of glucagon in the plasma and localization of the tumor with the aid of arteriography or exploratory laparotomy. Healing has been obtained and recognition from the dermatologic picture is mandatory. Steroids are sometimes effective.

Lightman and Bloom described a case of glucagon-induced diabetes mellitus (258).

The patient was a 50-year-old woman whose symptoms started with perineal irritation and Candida infection. The next year glucosuria was found. Her rash became worse and spread to fingers and shins with red, scaly, treatment-refractory lesions. The course was relapsing with weight loss and a smooth tongue with vermilion color. Anemia (hemoglobin, 11.4 g/100 ml) was noted. Chlorpropamide did not control her diabetes, and she was given increasing doses of insulin. She had no ketonuria. A pancreatic tumor was suspected, and a laparotomy was performed in March 1973. A tumor 4 cm in diameter was found and removed together with the tail and one-third of the head of the pancreas. After one week her tongue, hands, and groins were almost normal, and her insulin requirements decreased. The tumor contained increased amounts of glucagon. The plasma glucagon level was 20 times higher before the operation than it was postoperatively, when it was normal.

This is the first proof that glucagon can cause diabetes mellitus in man.

A paradigm of truly paraneoplastic symptoms that have shed new light on obscure problems was published by Gleeson et al (154).

A woman of 44 was admitted to Hammersmith Hospital for polyuria and nocturia lasting about a year. She had had severe constipation for many years. Lately she had developed generalized increasing edema and for some weeks she had vomited, had become amenorrheic, and had lost scalp hair. She had also noticed a skin rash. On admission a typical iron deficiency state and an increase in ESR to 90 mm/hr (Westergren) were noticed. She had severe hypoproteinemia. Steatorrhea was present but xylose absorption was normal. Her glucose tolerance test showed a mildly diabetic pattern with a normal fasting blood sugar. The Schilling test was abnormal both with and without intrinsic factor. On roentgenograms the small intestine was dilated with a coarse, thickened mucosa. The transit time was very rapid. Surgical biopsy of the jejunum was done. The specimen showed marked hypertrophy of the villi (easily visible to the naked eye). There was edema but no signs of inflammation. Happily for the patient she developed urinary tract infection and a pathological X-ray film indicating a renal tumor was found. Nephrectomy was performed. The tumor was well-defined but one small lymph node also contained tumor tissue. The histological pattern varied, but spherical electron-dense secretory granules were found in many cells. These granules stained strongly for tryptophan suggesting a carcinoid or a pancreatic alpha-2 cell tumor. Antiglucagon staining was strongly positive, and the patient's plasma contained more than 10 times the normal amount of immunoreactive glucagon.

The postoperative course was interesting. The constipation disappeared almost immediately and she had three to four stools a day. Hair growth and menstruation returned and the results of a biopsy of the jejunum became normal. She no longer showed a diabetic pattern of glucose handling.

Tumor extracts were prepared and their chemical nature analyzed. Bloom was able to prove that the tumor contained enteroglucagon (49).

In 1961 Unger found glucagon activity in the gut, and since then, the distribution in the gut has been studied in several animals. In 1974 Bloom published an excellent review on enteroglucagon (80). He points out that this hormone is different from the pancreatic hormone.

In 1976 Polak et al, from the Hammersmith Postgraduate Medical School, published the results of an interesting study, on the presence of the so-called pancreatic polypeptide in 33 patients with endocrine neoplasms of the pancreas. This is a well-defined substance with 36 amino acids that have been sequenced. Its functional role is still not defined. Fluorescent antisera were specific and showed beautiful staining of pancreatic polypeptide cells in glucagonomas and insulomas producing insulin as well as in normal pancreas. The sera from a number of patients showed increased levels of pancreatic polypeptide. It is possible that this will become an important screening method in suspected cases of endocrinologically active pancreatic tumors. (The reader should consult this interesting publication: "Pancreatic Polypeptide in Insulinomas, Gastrinomas, Vipomas, and Glucagonomas." Lancet 1:328, 1976.)

A. Enteroglucagon

Enteroglucagon is released by long-chain triglycerides and glucose. The elevation is very prolonged and does not decline until after five hours. It seems very probable that enteroglucagon has to do with dumping since such patients may have an enormous rise in enteroglucagon. These patients have an initial period of very rapid intestinal propulsion followed by a period of inhibited motility that may coincide with the enteroglucagon peak.

The renal tumor, that was investigated by Bloom, contained a substance that cross-reacted with some glucagon antibodies but not with an antibody specific for pancreatic glucagon (49). On elution from gel columns its position was identical to the peak of the hormone from fresh bowel extracts. According to Bloom, "It seems reasonable to hypothesize that enteroglucagon acts to inhibit motility, when unabsorbed food passes down the gut."

9 Disorders of Derepressed Immunoglobulin Synthesis

Each immunocyte clone produces one specific protein on challenge. At the same time this clone of cells proliferates and it must be assumed that multiplication of cells in the clone is linked to the stimulation of specific immunoglobulin synthesis. The normal challenge to produce immunoglobulins is contact with an antigen. This leads to a rapid increase in the titer of the corresponding antibody and to an increase in lymphocytes as long as high-molecular-weight antibody (19–20S) is produced and later to proliferation of plasma cells probably belonging to the same clone.

The cytological picture of macroglobulinemia is characterized by a great proliferation of lymphocytes but transitions to plasma cells are often seen. In myeloma, when immunoglobulins of the 7S type are formed, the plasma cells dominate the picture completely. There has been much discussion about the switch over from IgM to IgG antibodies against the same antigen. Recent studies on patients, who synthesize both monoclonal IgM and IgG, indicate very strongly that the variable (antigen specific) part of the immunoglobulin molecule is specific and constant, whereas the constant part of the molecule varies from IgM to IgG. The variable part is attached to the constant part first in the macroglobulin and then in the IgG. This should mean that the templates for the formation of the specific antibody site of the molecule are stimulated first and this leads to increased formation of the rest of the macroglobulin molecule including the light chain. At present studies are rather extensive and seem to prove that an identical light chain is attached to the heavy chain of either the IgM or the IgG molecule produced in this cell clone.

In previous publications I have drawn the parallel between the immunoglobulin-forming neoplasms, when there is no direct challenge with an antigen, and the tumors that produce topic or ectopic paraneoplastic polypeptides (464, 468). One problem in this discussion has been the fact that the immunoglobulin consists of several polypeptides that are coded by different genes (light chain, variable and constant part of heavy chain). We have reason to believe that each ectopic paraneoplastic situation is explained by the activation of one polypeptide-forming template. In all probability there may be a number of such activated templates for amino acid chains (polypeptides) in a carcinoma patient, but only one or a few have an activity that helps us to recognize them. The fact that a number of tumors producing more than one recognizable polypeptide has been found is of course not difficult to understand if the production is more or less random. On the other hand it seems clear that tumors arising from certain tissues, e.g., the so-called APUD cells, have a much larger "repertory" and are much more readily "derepressed," than others.

For more than 20 years I have tried to eradicate the term paraprotein. This should mean that the protein formed by plasma cells in myeloma or in benign gammopathy is not normal. It would be much easier to understand if normal templates were activated, and this seems to be the case. We started, with Winblad, systematic investigations of antibody activity in high titers in sera from patients with myeloma and found a few cases with very high antistreptolysin titer (462, 477). These cases were discussed in a Harvey lecture, 1961. Later we, and others, have found that there are many monoclonal immunoglobulins in the gammopathies that have definite antibody activity. It seems probable that this is explained by the fact that there is nonspecific activation of one immunoglobulin-forming template system. To me it is clear that nonspecific derepression of *one* polypeptide-forming template secondarily may lead to subsequent proliferation of the corresponding cell clone.

A number of these conditions have clinical symptoms and I have grouped them all under the general term "maladies of derepression".

In a paper from 1970 I mentioned as instances of such maladies of monoclonal derepression the following (468): (a) hyperviscosity, IgG and above all IgM; (b) cryoglobulins of different types; (c) cold agglutinins, IgM kappa; (d) cold urticaria; (e) warm hemolysin, IgG kappa or lambda; (f) Donath-Landsteiner, IgG lambda?; (g) anticoagulants? IgG kappa or lambda. Amyloid may be formed when the templates for light (chiefly lambda) chains are derepressed. Several types of kidney disease (Fanconi syndrome), as well as hypercalcemia, may be connected with such processes. (h) It is well known that proliferation of one immunocyte clone may lead to inhibition of other clones. Such patients may suffer from secondary Ig deficiencies.

A number of different clinical syndromes may thus develop as a consequence of monoclonal proliferation of immunoglobulins. In some instances the symptoms are caused by a special activity or by the increased amount of immunoglobulin in the blood. Examples of this are the cold agglutinin syndrome and the symptoms connected with cryoglobulinemia. In other instances the physical properties of the protein may have an unspecific effect, for example, by causing a hyperviscosity syndrome or disturbed renal function. A third group consists of patients who suffer from antibody deficiency because of inhibited synthesis of immunoglobulins other than those produced by the proliferating clone.

1. COLD AGGLUTININ SYNDROME

The most specific effect of a monoclonal immunoglobulin is exerted by macroglobulin molecules acting as cold agglutinins. This syndrome has been known for a very long time, and it was said that it occurred in a number of different diseases and was sometimes "idiopathic." It was regarded as one of the symptoms in acquired hemolytic anemia until it was established that the cold agglutination was the cause of the hyperhemolysis.

Investigations during the last decade have clarified many obscure points. It is now certain that there are two types of cold agglutinin disease, one reversible and another irreversible—or if you prefer the term "eternal." The first is seen in rare patients with excessively high titers of polyclonal cold agglutinin, usually after Mycoplasma infections but sometimes after other acute infectious diseases such as morbilli. This may lead to an acute severe hemolytic condition that has to be treated actively to keep the patient alive until the titers decline.

The other is connected with a monoclonal increase in macroglobulin and never disappears. In many cases the monoclonal fraction may be difficult to visualize because of its rather modest size. We have seen one such patient, who had cold agglutinin disease with no M-component visible on electrophoresis. On the other hand immunoelectrophoresis showed a definite component in the IgM fraction, and it was found that the M-component was "hiding" behind the transferrin band. Later its concentration increased and it was now visible on direct inspection.

According to our experience it is not unusual to find that these IgM components remain on the same modest scale for many years (see also 183). The condition must then be regarded as a parallel to benign monoclonal gammapathy of other kinds. In itself the proliferation of the clone is benign and nonprogressive, but it causes a lot of symptoms connected with the agglutinating activity of the immunoglobulin. It is difficult to decide how common such an active component may be in a population of patients with a proliferating macroglobulin producing clone. Patients with cold agglutinin disease come to the hospital because of their anemia and their protein disturbance is then diagnosed. It is not rare to find that some patients with an IgM component of monoclonal type have a slight cold agglutinin activity in this protein. According to our experience, there is a spectrum of activity in these different macroglobulins. Several investigators have tried to answer the question if the whole M-component consists of cold agglutinin. In most instances it is possible to adsorb the cold agglutinin with red cells in the cold. If this is carried out several times, tests show that almost the whole IgM band has disappeared. This is in accord with the homogeneity of the monoclonal globulins.

Nothing is known about the particular structure of the molecule that is responsible for the binding, but the work of Harboe in Oslo has demonstrated that the kappa type of light chain is—with a rare exception—present on the molecule (270). The reason for this interesting connection between light-chain type and immunological activities is not clear. So far it seems to be almost unique even though other predilections for lambda or kappa types are known.

In many patients with a cold agglutinin skin pallor is a striking sign. In others slight jaundice is also present. Rare patients, probably with very high titers or possibly with a temperature zone of activity that is unusually high, develop a syndrome of blue hands, ears, and nose. This blue face is unusual but very characteristic, and in the rather large number of cases I have seen, it has usually, but not always, indicated cold agglutinin and not cryoglobulin. Hemoglobinuria on cooling and violent exercise is rare. In a few patients the cold agglutinin is also a cryoglobulin that may be isolated by precipitation in the cold. This is very rare, however.

Treatment is very unrewarding. Steroids have been tried but with very little effect. Intense courses with alkylating agents have not been of much use in our experience but some successes have been published. Fudenberg has obtained good results with plasmapheresis.

It is important to remember that some patients may have a very stable level of macroglobulins with cold agglutinin effect.

We observed a woman born in 1873, from July 1959 to March 1969, when she died of coronary sclerosis and cardiac decompensation. During all these years she had had about the same level of IgM in the serum (± 10–15 g/l). Her cold agglutinin titers always had been extremely high (maximum, 1/325,000). She had a hyperhemolytic condition with low haptoglobin and a very high CO hemoglobin. Her anemia remained remarkably constant during the years in spite of the fact that she was not treated with cytostatics and initially was treated only for limited periods with small doses of prednisone (2.5 mg t.i.d.) for 3 months, then for another 2 months, and 5 years later for 5 years (463).

2. CRYOGLOBULINEMIA

The fact that serum globulins may precipitate in the cold was first established by Wintrobe and Buell (499). Some years later von Bonsdorff and Packalén ob-

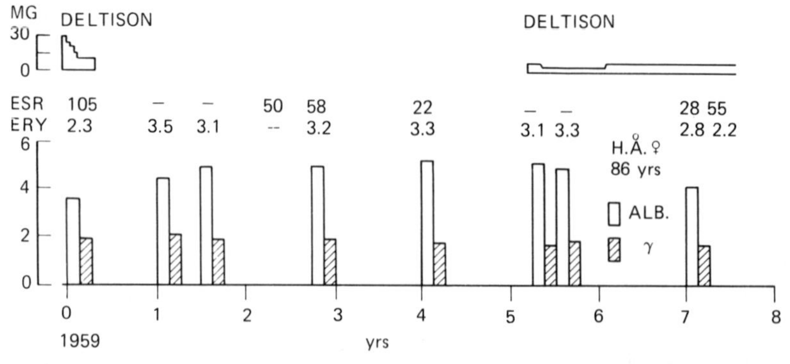

Figure 5. Serum protein pattern (lower row) in patient with very high titers of a macroglobulin (IgM) and of a cold agglutinin and with severe anemia that remained constant with very little treatment. No increase in IgM is noted ("benign" course).

served a similar case and proved that a gamma globulin was precipitating (53). In 1944 the present author published a report on the first cases of macroglobulinemia. In one of them the serum became a gel on cooling. Somewhat later Lerner and Watson observed a clinical syndrome that they called purpura cryoglobulinemia (247). Their patient showed precipitation in solid form of a serum globulin, and this phenomenon was investigated by them in different clinical conditions (cryoglobulinemia).

The literature on cryoglobulins has become very extensive. In 1952 I pointed out that the macroscopical appearance of such sera after cooling may be very different and gave relevant pictures (458). A small solid precipitate may form that should dissolve on warming. Several patients with vasculitis show a cold precipitable protein that does not dissolve on warming. According to definition this is not the true cryoglobulin, but the borderline between warm dissoluble and insoluble is very indistinct and the first type also has diagnostic importance. In other sera the formation of two layers may be seen, the lower gelatinous and the upper more limpid. Some of these sera have a very massive gel with a small upper zone. In some of these sera centrifugation at high speed brings down a solid precipitate. In other sera the insoluble protein is thixotropic and violent shaking may dissolve the gel. Under the microscope needlelike crystals sometimes can be seen.

It is not clear if the different physicochemical behavior of the protein is connected with special clinical pictures. The patient usually complains that he is very sensitive to cold, and that his hands become white or more often deep blue in cold weather. This is similar to what is seen in cold agglutinin disease even though I have rarely seen cyanosis of nose and ears in cryoglobulinemia. The condition described by Lerner and Watson as purpura cryoglobulinemia is rather variable in its manifestations (253). Sometimes true purpura appears. Many times the bleedings are more diffuse in the skin and necrosis of the skin of the finger tips and of other parts of the body has been noted after cooling. In some patients local cooling provokes an urticarialike erythema and transitions to cold urticaria may be gradual. Now it seems as if most authors would regard cold urticaria as really allergic and connected with specific reagin-type IgE molecules

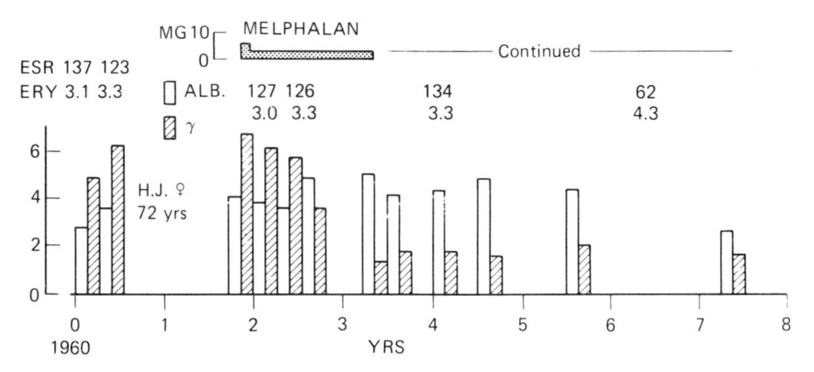

Figure 6. Development of serum protein pattern (lower row) in patient with cryomacroglobulinemia. Treatment with Alkeran also improved the anemia (possibly due to a change in dilution?).

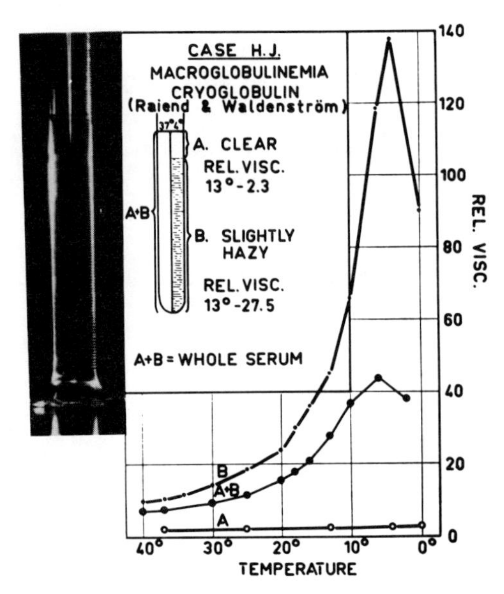

Figure 7. Data from the patient with macrocryoglobulinemia referred to in Figure 5. Two layers (A and B) develop on cooling the serum to 4°C. A is clear and much less viscous than B. The decrease in viscosity at 0°C is caused by precipitation of cryoglobulin. (From Waldenström in *Advances in Metabolic Disorders* with permission of Academic Press.)

and not with IgM. These borderlines should certainly be investigated more closely (92).

3. HYPERVISCOSITY SYNDROME

In my first publication on macroglobulinemia in 1944 it was noted that the serum was very viscous and determinations of relative viscosity with an Ostwald viscometer showed that the viscosity increased markedly with falling temperature. A special temperature viscosity index—the ratio between the relative serum viscosity at 13° C and at 37° C—was determined. An index above 120 was regarded as definitely pathological. One of the sera tested became so viscous at low temperature that it gelled, and the tube could be turned upside down without spillage (456). Later, the viscosities of a number of sera were measured, and it was found that this type of viscosity was common in clinical macroglobulinemia (457).

Schwab and Fahey (384) and Solomon and Fahey (407) investigated the part played by hyperviscosity in the development of different symptoms in macroglobulinemia (132). These authors performed plasmapheresis on a large scale removing as much as 2–6 liters of plasma in a rather short period. After this initial very massive removal of macroglobulins, the condition could be controlled by removing smaller amounts of plasma, e.g., 250–500 ml every week or biweekly.

The role of hyperviscosity in the production of clinical symptoms became well established from these experiments. The mechanism behind these changes remained obscure, however. For the diagnosis of hyperviscosity several observations are important. On ophthalmoscopy the eyegrounds are found to be markedly pathological. Venous dilatation, hemorrhages, and sometimes papilledema are seen. Greatly engorged veins are typical. Ophthalmologists, who have seen some such patients, maintain that they can make the diagnosis of a hyperviscosity syndrome from looking at the fundus. These changes may lead to serious disturbances of vision. The bleedings may cause progressive loss of vision, and instances of an increase in intraoccular pressure have been seen. Normalization of the viscosity leads to a disappearance of papilledema and to a decrease in venous size. Exudates, when present, may also disappear. These changes seem to occur pari passu with the decrease in viscosity (463).

Clinically the first signs of increased viscosity are often bleedings from the mucous membranes. Nosebleeds and bleeding gums are early symptoms. Purpura connected with thrombocytopenia is usually a late symptom and does not seem to be caused by the high viscosity. We have seen one man with purpura that greatly resembled the purpura hyperglobulinemia described in polyclonal hypergammaglobulinemia. It has now become clear that this type of purpura is caused by complexes of immune globulins and is usually accompanied by a very high titer of rheumatoid factors. This was the case in the patient with chronic purpura on the legs and macroglobulinemia. Such cases are rare, however, and I have seen only one other patient of this kind, who had a monoclonal IgM and an enormous increase in rheumatoid factor titers.

Patients who have very high viscosity and bleedings from the nose and gums may have normal menstrual bleedings. This was the case with the patient, whose clinical course is shown in Figure 8. At that time we had not yet started plasmapheresis, but the patient was treated very actively with cytostatics and steroids, and there was marked reduction in the macroglobulin content of the plasma over the years. At the same time the relative viscosity decreased from 22 to 3.7 at 0°C. The condition of her fundi became much improved and the eyesight in one eye was saved. Her bleedings improved.

A number of patients with severe hyperviscosity develop a condition that resembles a coma. We have seen some patients with macroglobulinemia and one patient with myeloma who obviously had this type of disturbance. In the latter case very massive plasmapheresis was performed and the patient's condition improved remarkably.

A priori it seems tempting to assume that patients who have a very viscous plasma, which obviously causes high circulatory resistance, should suffer from cardiac decompensation. This result seems to be quite rare, however, and Solomon, who has seen many cases with this disturbance, reports on only one case with marked cardiac decompensation that was much improved after massive plasmapheresis. Other symptoms occurring in macroglobulinemia but obviously not connected with hyperviscosity are discussed later.

Initially plasmapheresis was performed in plastic bags with spontaneous sedimentation of the red cells and removal of the plasma. This was a very time-consuming procedure. The development of the cell centrifuge for the whole blood made it possible to remove large quantities of different blood constituents easily. We have seen a number of patients, who have profited from this treatment. It should be remembered that macroglobulins may have some rather disagreeable physical qualities. On dilution they precipitate and may cause block-

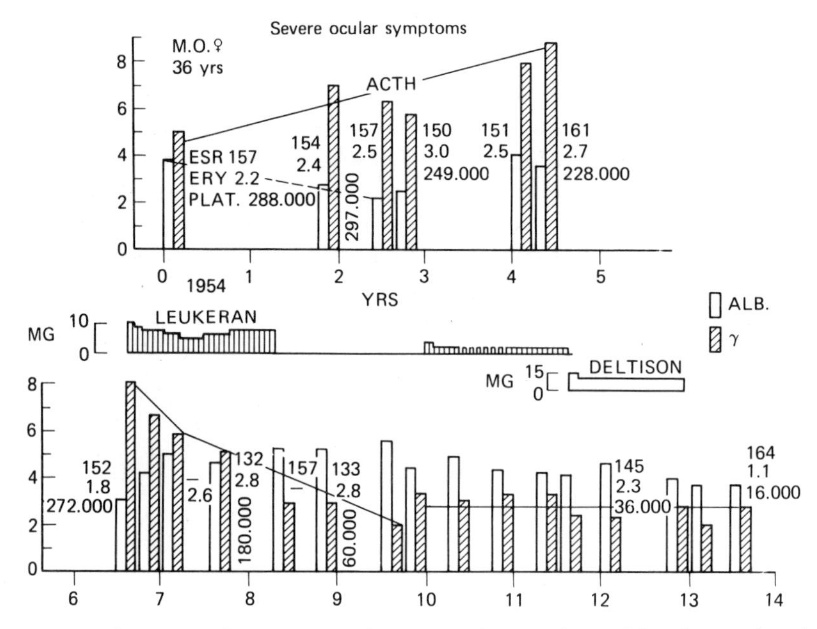

Figure 8. Development of serum protein pattern in a patient with a hyperviscosity syndrome during treatment with Leukeran.

ing of the tubes in automatic analyzers. The fact that some macroglobulinemic sera become very sticky at room temperature might cause trouble in the centrifuge. Therefore, the relative viscosity at room temperature always should be determined before centrifugation is started.

4. HEMOLYSIS

Hemolysins belonging to the immunoglobulins and active at body temperature ("warm") seem to be polyclonal. Hemolysis has been investigated extensively in recent years with the aid of antibodies developed against the light-chain component of the immunoglobulin molecule. In some patients it has been possible to inhibit the hemolytic effect with either kappa or lambda antibodies, but the results usually were not very clear-cut. So far there has been no report of a myeloma-produced warm hemolysin, and it is improbable that such a case should escape notice.

The most striking among the immunological hemolytic mechanisms is without doubt the Donath-Landsteiner type of cold hemolysin. It has long been known that this is in some way connected with the Wassermann reaction, even though its dependence on syphilitic infection has not been proved conclusively. This type of hemolysis has become extremely rare, probably due to the fact that active treatment of syphilis has improved. In a recent case from Israel the immunoglobulin was very carefully examined immunologically, and it was established that the hemolysin was pure IgG lambda; confirmations of this observation are needed, however. Bastrup-Madsen et al observed a patient, who had a tendency

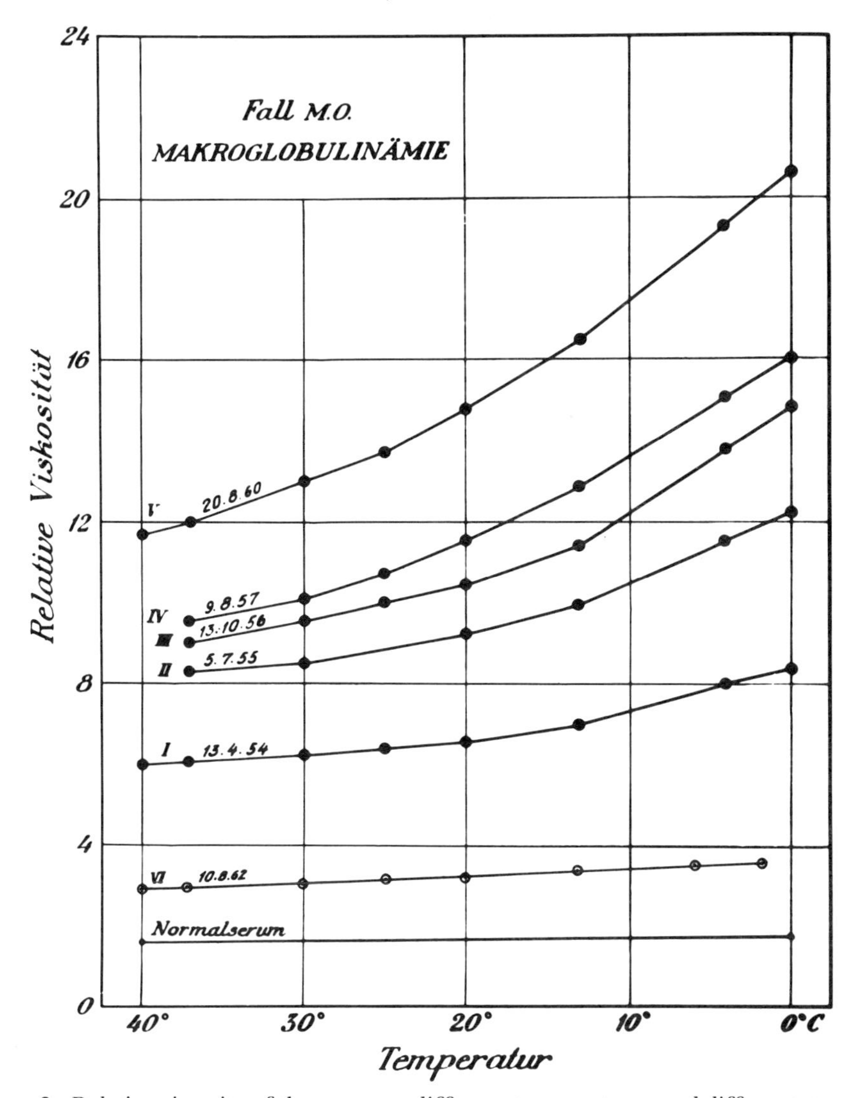

Figure 9. Relative viscosity of the serum at different temperatures and different stages of the disease in the patient referred to in Figure 8. Lines I–V show a continued increase in viscosity before treatment, whereas line VI shows a marked decrease after successful treatment with Leukeran. In this patient treatment with cytostatics had the same effect on plasma viscosity as plasmapheresis even though it was very slow. Initial rapid improvement should be obtained by plasmapheresis and should be stabilized by cytostatics. (Waldenström and Raiend unpublished.)

to form urticaria after exposure to cold (24). He later started having typical attacks of cold hemoglobinemia. A typical Donath-Landsteiner type of haemolysin was demonstrated, and it was found that a monoclonal IgG was the reagin. The patient was followed for 6.5 years and the M-component did not increase. These are regarded as the only well-documented examples of D-L reaction connected with an M-component.

5. MANIFEST ACTIVITIES OF DIFFERENT MONOCLONAL IMMUNOGLOBINS

I always have maintained that the word "paraprotein" should be avoided since it should indicate that the proteins formed by tumor cells are pathological like the cell itself. In my opinion it is much more plausible to regard these proteins as normal globulins produced in excess. To determine which view is correct, in 1953 we started a program, in Malmö in collaboration with the departments of medicine and clinical immunology, to study the patterns of polyclonal and monoclonal hypergammaglobulinemia sera. Professor Winblad and I set up a battery of tests for the study. This consisted of the Wassermann, Kline, and Meinicke reactions for syphilis, a test for anticomplementary activity of sera warmed to 56° C, a number of agglutination tests for different bacteria (typhoid and paratyphoid, bacillus abortus Bang, tularemia, and later Listerella and Yersinia). The presence of heterophil antibodies (Paul-Bunnel) and cold-agglutinins was also determined. Initially only the Waaler-Rose test for rheumatoid factors was used, but later some tests for antihuman gammaglobulin were added.

Winblad and I published a preliminary report of the study. Sera from 12 patients with myeloma were investigated along with sera from patients with broad-banded (polyclonal) hyperglobulinemia. We continued the investigation and Winblad, Hällén, Liungman, and I published the results for monoclonal hyperglobulinemia (myeloma, benign, gammapathy, and macroglobulinemia) in 1964 (477).

The chief problem was to find if there were indications of specific activities of an M-component. In three patients with a very big IgG M-component extremely high titers of antistreptolysin were found. A cold agglutinin activity was found only in some sera from patients with macroglobulinemia. This was also true of the rheumatoid factors that were present in very high titers in some sera from macroglobulinemia patients without any signs of rheumatic disease. One of these patients had a type of gamma globulin aggregation that improved spontaneously. We concluded that in the proliferation of one clone of plasma cells the daughter cells maintained the formation of gamma globulin molecules with the original immunological pattern. Occasionally this may result in an enormous amount of one "specific" antibody in sera from a myeloma patient without a corresponding inducing antigen.

In 1957 Christensen and Dacie demonstrated that cold agglutinin disease was caused by a monoclonal macroglobulin. Fudenberg and Kundel published the findings of investigations of a monoclonal cold agglutin IgM. In 1961 Kritzman and Kunkel described a Waldenströmtype macroglobulin with rheumatoid factor properties; and in the Harvey lectures that year I discussed some IgG sera with excessive titers of antistreptolysin.

A number of observations of false-positive serological reactions for syphilis have been described in connection with monoclonal immunoglobulins. Cooper et al observed a patient who had clinical signs of lymphoma and a big M-component of IgM kappa type (87). The serum showed VDRL reactions positive at 1:4096 dilution, whereas the TPI was negative. Incubation of the serum with mercaptoethanol reduced the titer to 1:64. Testing against different antigens gave a precipitin line with cardiolipins and lecithin but not with cholesterol. Isolation experiments showed that the purified macroglobulins gave a positive precipitin reaction with nine of 15 different phospholipids tested. Lysolecithin gave no precipitation nor did low-density lipoproteins.

Killander et al found a patient, who had a positive Wassermann reaction but a negative Kline and TPI test (233). The reagin was a macroglobulin M-component. Lecithin was probably the antigen. It seems probable that several cases of antithromboplastin activity causing coagulation disorders may belong to the same group, although this is much more common in the polyclonal type of gamma globulin increase seen, for example, in patients with SLE.

In 1974 Drusin et al described a 73-year-old man with clinical macroglobulinemia. His serum showed a false-positive Wassermann reaction 38 years before the diagnosis of macroglobulinemia was made. On different occasions the titers varied from 1:8192 to 1:32,000. These authors stressed the point that false-positive reactions should be looked for in a large number of macroglobulin sera. (This has been done at Malmö without finding many positive results.) The case published by Drusin et al demonstrates in a very convincing way that macroglobulinemia may be present long before it is finally diagnosed.

Osserman has seen at least three patients in approximately 100 with increased IgM levels and strongly false-positive reactions.

One patient, an 80-year-old woman, had had a biologically false-positive test for syphilis for 22 years, when she was admitted to the hospital with splenomegaly and severe anemia. Serum protein electrophoresis showed a monoclonal IgM band of 2.5 g/L. Kline, Kolmer, and VDRL tests were positive and the FTA and other specific tests for treponema were negative. Her serum contained a panagglutinin but no cold agglutinin. Rheumatoid factor activity was demonstrated with the bentonite flocuulation test. After treatment with large doses of prednisone, she had an excellent remission with marked reduction of IgM. The light chain type was kappa but she had no Bence Jones protein in her urine. The isolated IgM was thought to be only partly responsible for the different abnormal serological reactions, but the complicated findings are difficult to interpret.

In 1960 David Miller described a case of Waldenström's macroglobulinemia in which the macroglobulin was precipitated by heparin (294). The macroglobulin was isolated as a euglobulin by repeated dilutions. The precipitate was dissolved to obtain a protein concentration of 3 g/100 ml. The dissolved protein seemed to be homogeneous at ultracentrifugation but as usual peaks were found at 7.5S and 27S. Zone electrophoresis showed a homogeneous peak. When heparin was added in different concentrations, the macroglobulin precipitated. This had been observed on a sample of heparinized blood that showed an intermediate gelatinous layer at centrifugation. It was difficult to decide if this was a real antigen-antibody reaction or if it represented complex formation of some other kind. The precipitation occurred in a very broad zone of different heparin concentrations. Chondroitin sulphate gave no precipitate. The practical conclusion was that heparin should be given with extreme caution to such patients with macroglobulinemia.

A similar complication was observed by Harboe. (The patient was first described by K. Bauer [25].) After the divalent X-ray contrast medium io-glycamic acid (Biligram) was given to the patient, he went into shock suddenly and died, obviously because of precipitation of the macroglobulin in his circulatory system. The mechanism was further analyzed in Harboe's laboratory, and it was established that the precipitation depends on the formation of a lattice between antibody molecules and an antigen that has to be divalent. The monovalent compound reacted with the globulin molecule but caused no precipitation. It inhibited the precipitation caused by the divalent substance. It is probable that other similar substances are also precipitated with monoclonal macroglobulin. (Type-specific Klebsiella polysaccarides are precipitated in this way.) This mechanism is probably very rare as a serious complication in clinical medicine. No other macroglobulins from sera from a number of different patients precipitated with Biligram.

In 1972 German authors published interesting data on a patient who obviously produced an IgG-forming complex with transferrin. The patient was an elderly lady with an excessively high serum iron concentration (750 mg/100 ml) and signs of pigment cirrhosis. The transferrin was obviously bound to the Fab part of a monoclonal IgG kappa, probably on a 1:1 basis. So far this is an unique "autoantibody" (489).

Recently Farhangi and Osserman have published the findings of very extensive investigations regarding a patient who showed yellow skin and hair but no jaundice of the sclerae (136). This patient had a monoclonal IgG lambda in her serum and showed typical symptoms of myeloma. Her bilirubin was normal, but it was found that the M-component on precipitation was yellow, and it was possible to prove that riboflavin was specifically bound to the Fab part of the molecule. Other related possible haptens were also investigated for binding. It was found that a large number of other monoclonal Ig molecules did not bind riboflavin. The authors point out the fact that this unique property of the M-component was detected only because the hapten was colored.

A very high antistreptolysin activity has been observed in sera containing M-components. As a matter of fact this seems to be the most common activity found in an M-component of IgG type. Mansa and Kjems found very high antistaphylolysin activity in an IgG M-component, and we have seen similar activity in the serum of a patient with myeloma (see 451a). Zettervall analyzed activities of two M-components with antistreptolysin activity and serum from another patient with myeloma with very high antistaphylolysin activity. (For further information, the reader should consult Zettervall [509].)

A number of investigations of monoclonal immunoglobulins with activities as hemolysins or as anticoagulants have given results that are hard to evaluate. Castaldi's and Penny's article contains interesting information on this subject (76).

Riesen et al screened a large number of sera containing M-components for antibodies against low-density lipoprotein (LDL) (354). It is well known that patients who have had multiple transfusions may develop such antibodies. In two IgG myeloma proteins the authors were able to show that the binding site was on the Fab fragment. Precipitation against delipidated LDL could be observed, whereas no precipitation occurred with the apoprotein in high-density lipoprotein (HDL). It is probable that the M-components reacting with both LDL and HDL as described by Beaumont are directed against the lipids. Their real effect is very hard to evaluate, however.

Videback et al found a patient with myeloma who had an IgA kappa protein. The antistreptococcal hyaluronidase (ASH) activity in the serum was much higher than had ever been observed before (5.2 mill. U/ml). The activity was connected with the Fab fragment. It is thus clear that even this specificity may be present in myeloma globulins even if the antistreptolysin-containing sera do not show ASH activity (451a).

A patient with myeloma and marked hypercalcemia, who did not show any obvious clinical signs of this metabolic disturbance, was recently observed in Malmö. The monoclonal IgG found in the patient's serum was investigated in some detail by Lindgärde et al, and it was found that calcium ions were bound specifically to the immunoglobulin (258, 259). It was shown that the binding was localized to the Fab fragment, and that the light chains were active in calcium binding. These findings are compatible with the assumption that calcium is bound in the antibody-combining site.

An interesting observation of a copper-binding immunoglobulin IgG lambda and with restricted Gm activity was published recently by Baker and Hultquist. It was clear that copper was bound only to the Fab fragment. Two atoms of copper were bound per molecule. Further investigations of this myeloma globulin must be expected. (Unified FACEB 1976.)

Potter has been able to show that a number of mice with plasma cell tumors may produce immunoglobulins against phosphorylcholine. These reactions may be regarded as authentic antigen-antibody reactions. In a patient with macroglobulinemia the M-component was investigated for specific activities. It was then found that binding activity against phosphorylcholine was very high and could be used for affinity chromatography with this substance as the immunoadsorbent. Only one serum out of 900 with M-components reacted in this way. The characteristics of binding agreed with the true antigen-antibody mechanism. The IgM was of kappa type.

The real connection between appearance of monoclonal globulins and a number of different clinical conditions is not well established. It has been known for a long time that pyoderma gangrenosum is connected with increase in IgA and lichen myxedematosus, chiefly with IgG (469). It is not clear if these patients have myeloma or if they have developed a monoclonal antibody that is active against some constituent of the skin. In the latter instance this would be an "autoimmune" disease caused by the proliferation of one plasma cell clone (see Chapter 1).

In general oncology it is not uncommon to find carcinoma patients who suffer from the production of specific polypeptides with hormone activity. These paraneoplastic phenomena most likely may be compared with random formation of immunoglobulins by special clones of plasma cells or lymphocytes. It must be remembered, however, that the normal function of these latter cells is to synthetize immunoglobulin. The product is therefore topic, but the derepression is active only in one clone of cells. In many carcinomas the paraneoplastic product is completely ectopic, whereas it is clear that templates for the formation of *one specific* polypeptide are activated in such cases. I would regard all these conditions, both topical and ectopic, as caused by derepression of polypeptide-forming systems with continued production without further challenge.

The fact that "autoimmune" diseases are very rare in connection with myeloma may speak against the assumption that the presence of one specific antibody molecule may initiate the development of dermatomyositis or chronic thyroiditis.

10
Neoplasias Producing Specific Substances

1. Carcinoid Tumors (Argentaffinomas)
2. Medullary Thyroid Carcinoma (MTC)
3. Mastocytoma (Mastocytosis)
4. Pheochromocytoma, Neuroblastoma, and Ganglioneuroma

1. CARCINOID TUMORS (ARGENTAFFINOMAS)

Kultschitzky described a special type of cell in 1897 that occurred isolated in the intestinal epithelium near the lamina propria. Later authors found that these cells were granulated and chromaffin and also had the capacity to reduce silver salts. They have been studied extensively by many pathologists, and it was thought that the granules were connected with the secretion of some specific substance. A German pathologist, Feyerter, became very interested in this problem and wrote extensively about a special organ that he called a "peripheral endocrine gland" (138). Because of its tinctorial characteristics he called the organ "das helle Zellenorgan," the light cell organ, and studied the occurrence of these cells in a large number of different organs besides those of the gastrointestinal tract. In 1939 Erspamer published the results from pharmacological studies of the enterochromaffin tissue. He had extracted a very potent substance that he called enteramine. This was later identified with serotonin and the chemical constitution was determined as 5-hydroxytryptamine (5-HT). Thus these cells produce one of the potent biogenic amines. Gastric carcinoids may also produce histamine which causes a peculiar flush (459, see also 476 and Figure 1).

These argentaffine cells may form tumors in the intestinal wall called argentaffinomas or, more commonly, carcinoids. (The latter term has more or less superseded the more correct one.) These tumors were first regarded as usually benign and quite small with very few clinical symptoms, but instances of malignancy were described with multiple metastases in lymph glands and liver. In 1953 we observed two patients who had malignant carcinoid tumors with massive metastases and who also suffered from a syndrome with flushing, diarrhea, and wheezing respiration occurring in attacks (473, 477). This syndrome was studied extensively and we noticed the presence of right-sided heart disease in two patients (439). At the same time Isler and Hedinger published their postmortem findings in three cases with right-sided valvular heart disease and very marked thickening of the endocardium on the right side of the heart (215). This syndrome was studied in great detail during the following years. Together with Pernow (334) we were able to demonstrate the presence of 5-HT in increased amounts in blood and urine from several of our patients (466). At the same time Udenfriend et al and Sjoerdsma et al studied the metabolism of 5-HT and found that it was formed from 5-hydroxytryptophan and broken down to 5-HIAA. This is an important metabolite for initial diagnosis and also for control of therapeutic results (400, 448).

The typical syndrome in these patients is only seen when considerable amounts of tumor tissue are present. Thorson has pointed out that it is necessary to have the primary tumor or its metastases located with direct venous outflow into the heart (437). This means that intestinal carcinoids with liver metastases or carcinoids in ovarian teratomas draining blood into the inferior vena cava and bronchial carcinoids of a certain size should be suspected when the clinical syndrome is present. The presence of lymph gland metastases in the abdomen is insufficient for the development of the syndrome.

Therapy in metastasizing carcinoid tumors is not very effective. The diarrhea may be improved by giving parachlorophenylalanine which inhibits the synthesis of 5-HT. Surgical therapy is usually technically impossible. We have seen one patient with the complete syndrome who had a very large metastasis in the right lobe of the liver. This was resected and after a very stormy postoperative phase the patient's condition improved greatly, and he is in rather good health 10 years after the operation (9).

In addition to flushing and diarrhea, other clinical symptoms may be of interest. During the flush attack it is common to find severe hypermotility of the gut with borborygmus and sometimes colics. Some patients also show wheezing and rhonchi at the same time. Such asthmatic attacks are rare, however, and never cause serious trouble. They are probably caused by the same active substance. The prognosis in this malignant condition is very unusual. Many patients have been followed for one or two decades with metastases to the liver that were found to be widespread and massive at an early exploratory laparotomy. Other patients had pulmonary metastases for a long time without much inconvenience. There are, however, patients with a more malignant course, but the majority have a fairly good prognosis *quoad vitam*. The final stage is usually heralded by increasing cardiac decompensation and *right-sided heart disease*. The earliest symptoms seem to arise from the pulmonary valves with stenosis as the dominating symptom. Incompetence in this valve also develops. The most characteristic symptoms occur, when the tricuspid valve has become incompetent. In the final stage many patients have expansile pulsations of the liver and systolic pulsations of the neck veins indicating the presence of this valvular lesion. In these patients the diagnosis is easy but in many instances cardiac catheterization is necessary to obtain a firm clinical diagnosis. The development in a large number of our patients has been followed by catheterization and the results have been published (444). We have tried to obtain information from intracardiac biopsy regarding the presence of fibrosis in the endocardium in these patients but so far have been unsuccessful.

Physiologists have noted for a long time that serotonin is inactivated by passage through the lungs, and this has been regarded as the explanation of why

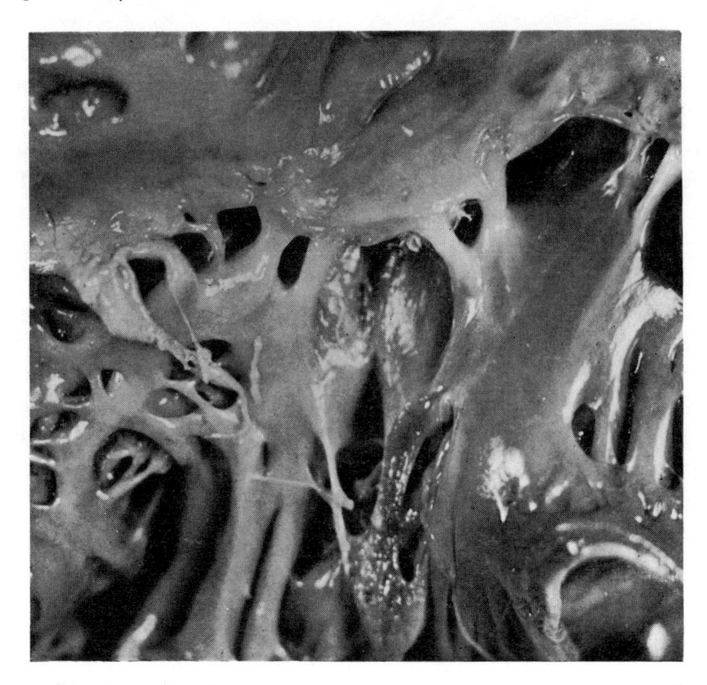

Figure 10. Thickened and contracted tricuspid valves and chordae in a patient with carcinoidosis.

only the right side of the heart is afflicted. An excellent confirmation of this theory was obtained by von Bernheimer et al (40). They observed a patient with a tumor of the lung that was resected. It was found that the pulmonary veins draining the carcinoid tumor had unusually thick walls. The patient died sometime after the operation, and the postmortem examination disclosed valvular lesions on the left side of the heart but not on the right. The authors conclude that this must be explained by assuming that there is a high level of the active substance in the blood coming from the tumor in the lung. The mechanism for this thickening and shrinking of the endocardium is not clear, but it is evident that not only the valves and other parts of the endocardium but also the peritoneum is influenced by products from the tumor. Isler and Hedinger noted in their first case report that the tumor in the small intestine was surrounded by fibrous masses, and in one case, they describe the abdominal status as resembling a "cast in plaster of Paris" (215). Surgeons know that long-standing carcinoids may cause severe technical complications, and we have seen some patients in whom it was impossible to perform operations.

I have already mentioned that the final stage of carcinoidosis is often characterized by increasing cardiac decompensation. Among the 11 patients that we have followed closely, right-sided cardiac failure has progressed even after radical surgery in two patients with ovarian teratoma. Our patients have had normal pulmonary venous pressures, but a hyperkinetic condition is obviously present in many patients during flushing. It has been postulated that this should be mediated via some kinin. The ballistocardiogram becomes markedly pathologic if taken during a flush and compared with the non-flushing condition in the same patient (440).

This situation deserves more attention. The two patients who underwent radical surgery showed no increase in urinary HIAA levels after the operation and the final postmortem disclosed no metastases. In spite of this they had progressive cardiac symptoms and died with decompensation (one 10 years and the other 9 years after surgery). The postmortem examination disclosed no metastases. It is probable that there was considerable transitory improvement of cardiac function but also continued shrinking despite relief from the toxic chemical influence (see also 9).

The connection between biochemical and clinical findings in this condition is illustrated in Figure 13. From the beginning we assumed that 5-HT was the cause of all these symptoms. It is very probable that the diarrhea is connected with 5-HT, and it is tempting to assume that the normal presence of isolated Kultschitzky cells along the wall of the intestine may be connected with normal peristalsis through local formation of 5-HT. Direct proof of this effect has not been available, but the results of parachlorophenylalanine therapy on diarrhea is striking. This drug was first used as an inhibitor of 5-HT formation, and we have been able to confirm that the diarrhea is much improved when the excretion of urinary HIAA is found to be reduced (126). The best proof that the drug is very effective is perhaps the fact that our patients ask for more when their tablets have been consumed (426). However, it does not seem wise to prescribe the drug for long periods because of the possible risk of interfering with 5-HT-production in the brain. In spite of very elaborate and numerous animal experiments, no one has been able to prove that the stimulation of connective tissue formation seen in these patients is mediated via 5-HT. It is also possible that some as yet unknown substance may be active in this connection (see also 168).

The role of 5-HT in the production of bronchial constriction is not definitely established. There is, however, another type of symptom that may well be con-

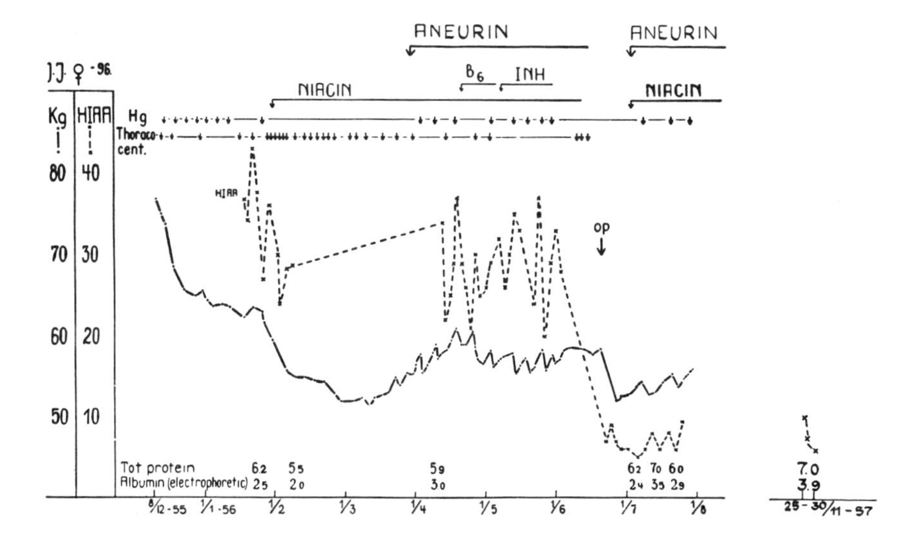

Figure 11. Clinical and biochemical data from a patient with ovarian teratoma containing argentaffine tissue before and after operation. It is seen that the levels of HIAA become normal postoperatively indicating radical extirpation of carcinoid tissue. (From Thorson [437] with permission of the author and of *Acta Med. Scand.)*

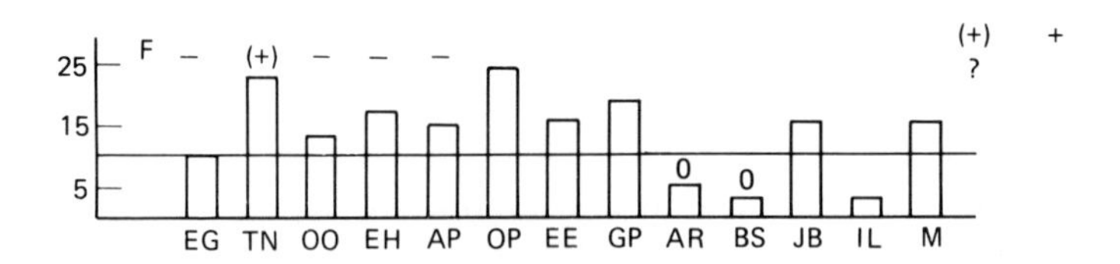

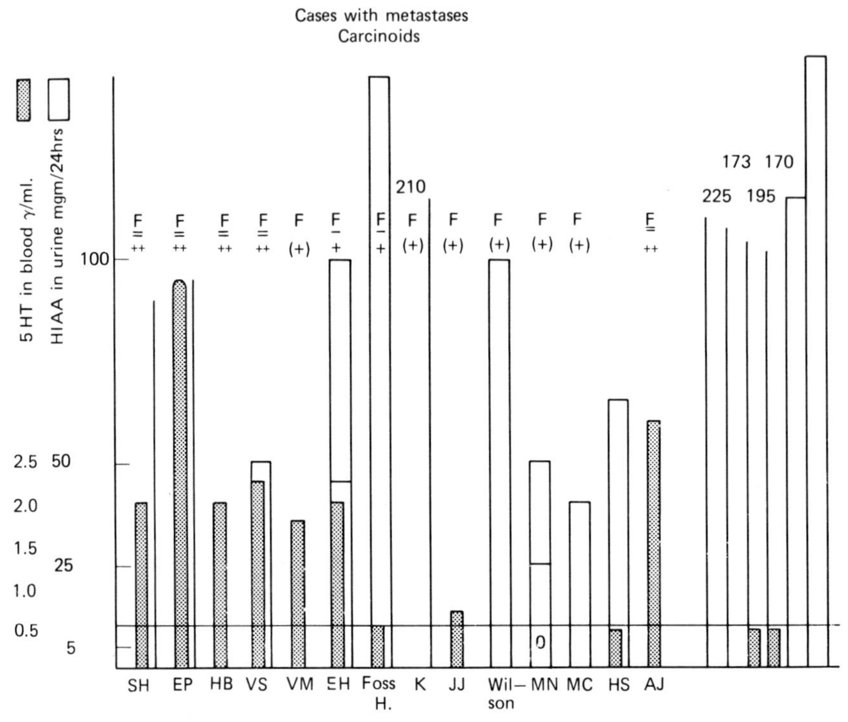

Figure 12 In patients who have undergone radical surgery, the excretion of HIAA appears to be normal (upper row), whereas urinary HIAA values are high in those patients who have had a radical operation It should be noted, however, that a few of the latter patients have normal values. This symptom therefore can not be regarded as a completely reliable sign of radical treatment. In a number of patients increase in 5-hydroxytryptamin was found in the blood. *F*, flushing. (From Pernow and Waldenström [475] with permission of the authors and of *American Journal of Medicine.*)

nected with the derangement of the outflow from the tryptophan pool as pictured in Figure 13. We have reason to believe that niacin synthesis may suffer when too much tryptophan is diverted into the formation of 5-HT. This mechanism was first discussed by Sjoerdsma, Weissbach, and Udenfriend (400). We have seen one or two patients with a dermatosis that may be compared to pellagra with rapid improvement after radical operation. To my knowledge there have been no quantitative determinations of the enzyme or attempts to cure the condition by giving large doses of niacin without other treatment. Our patients have not had the typical glossitis, and the diarrhea is certainly not caused by pellagra.

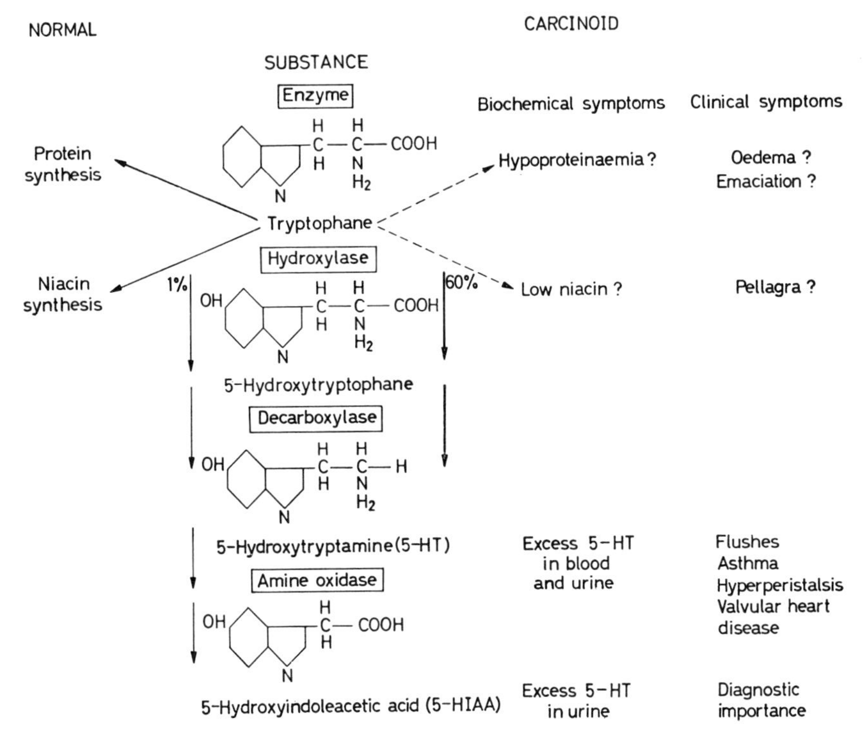

Figure 13. Clinical and biochemical data from a patient with ovarian teratoma containing argentaffine tissue before and after operation. It is seen that the levels of HIAA become normal postoperatively indicating radical extirpation of carcinoid tissue. (From Waldenström [460] modified from Udenfriend et al and with permission of Butterworth.)

The flush has been discussed in Chapter 1. It is probable that bradykinin or another kinin is the trigger substance. Very important work by the groups at St. Mary's Hospital in London and at the NIH in the USA has shown that catecholamines produce a flush if injected intravenously into carcinoid patients. We have repeated these experiments and confirm the finding that there is a very rapid decrease in blood pressure together with a flush and then a rise as an effect of the catecholamine (426). Many observers have noted that psychological factors, such as anxiety, start a flush. As a matter of fact this chain of events must start in the brain, but it would not be mediated via nervous impulses to the carcinoid tissue since tumor cells have no innervation. It is therefore probable that the chain includes the splanchnic nerves to the adrenal medulla. The last step is represented by catecholamines working on the tumor cells. Even though many facts favor the hypothesis that bradykinin is the active flush substance, there are still some facts that are difficult to explain.

2. MEDULLARY THYROID CARCINOMA (MTC)

In 1959 Hazard et al described a special type of thyroid carcinoma that had not been recognized earlier despite the fact that it has an interesting paraneoplastic picture and striking histological characteristics. It is a solid carcinoma and histologically it does not look well differentiated. There are no signs of follicular differentiation and, most important, the stroma contains large amounts of amyloid. In 1966 E. D. Williams suggested that this tumor might contain calcitonin and be composed of cells corresponding to the so-called C-cells of the normal thyroid (494). These had been shown to produce calcitonin and are usually supposed to derive from the ultimobranchial body. In 1961 it was found that this type of tumor may occur together with—sometimes bilateral—pheochromocytoma. The first author who made these observations was Sipple, who could also show that this combination of different tumors occurred in families. The syndrome has been named after him, and large pedigrees have now been studied. The biggest and best studied family originated in Sweden, but one female member emigrated to the USA in 1901 where she had eight children, seven of whom were carriers of the trait. Together with the Swedish family there are now a large number of known affected members (290). As in other families the majority of the patients have only thyroid carcinoma, but in rare instances, others may have only a tumor of the adrenal medulla or, more commonly, a combination of the two.

In recent years it has become possible to make an early diagnosis of the condition from determinations of serum calcitonin levels with the aid of radioimmunoassay. It has been found that patients who do not have any clinical symptoms of the disease may have elevated calcitonin levels. It is not yet known if this is an early sign of later tumor development or if the condition may remain benign for a long time. Calcitonin assay is also valuable for the determination of radical treatment (163).

Three separate syndromes with multiple endocrine neoplasia are known. Type I involves pancreatic islets, adrenal cortex, parathyroid glands, and pituitary. Another (type II) has medullary thyroid carcinoma as the chief disturbance. This may be complicated by pheochromocytoma (see 135)—often bilateral—and chief-cell hyperplasia or adenoma of the parathyroid glands (345). The latter finding does not seem to be a convincing example of multiple neoplasia since it would be much more logical to regard it as a compensation against the influence of thyroid calcitonin. Recently a third type has been distinguished. In type III the individuals have medullary thyroid carcinoma and pheochromocytoma, and multiple mucosal neuromas as a third neoplasia (316). Single type III cases do not appear in so-called type II families. In the so-called type III essentially all individuals have mucosal neuromata (307).

Baylin et al have produced good evidence that the inherited form of MTC is of clonal origin (28).

Determinations of metabolites in the urine are also of great diagnostic importance for tumors of the adrenal medulla. Members of families with Sipple syndrome should be checked for urinary metabolites.

The tumor tissue in the thyroid usually occurs on both sides. This may of course be interpreted as "metastases," but it is also possible that these cells are under the influence of some "tropins" that induce proliferation of different cells originating from the neural crest. This would fit in well with the fact that the pheochromocytoma is very often bilateral. The same situation is present in another endocrinologically active tumor with the same origin, namely argentaffinoma (carcinoidosis). There have been lively discussions as to whether the latter disease is systemic or metastatic. Similar discussions are likely to occur regarding mastocytosis which is also characterized by a proliferation of one cell type without destructive growth.

The fact that calcitonin is the most constant product in medullary thyroid carcinoma (MTC) might suggest that one of the paraneoplastic symptoms in this disease should be tetany clinically and hypocalcemia biochemically. Neither assumption is true. These patients are normocalcemic with very few exceptions. This has been explained by the assumption that there is a corresponding increase in parathormone production. The fact that the parathyroid glands are usually of normal size is perhaps difficult to understand. Secondary hypercalcemia or adenoma of the parathyroids is quite rare. Most authors seem to regard the production of calcitonin as having no biochemical importance, and none of the different clinical symptoms has been connected with the secretion of this hormone.

There are, however, two findings that may be caused by overproduction of calcitonin. One of the characteristic microscopical features of the thyroid tumor is that it contains amyloid. This is always a completely localized thyroid symptom, and generalized amyloidosis is never seen. It is also known that the tumor in MTC contains very high levels of calcitonin. This hormone is a polypeptide with a molecular weight of 3600 daltons and consists of 33 amino acids. In a recent discussion on amyloid, I suggested that the amyloid may consist of calcitonin fibrils arranged in such a way that they behave like amyloid (484). Recent investigations seem to have proved that this is correct. As a matter of fact calcitonin dimers with antiparallel arrangement of polypeptide chains are known. This is a prerequisite for the formation of amyloidlike material.

The fact that calcitonin has an excellent therapeutic effect in Paget's disease has been explained by stimulation of osteoblasts and a consecutive decrease in osteoclasts. Possibly, transformation of the latter cells to the former is induced. It is well known that tumors of MTC may show calcification both in the glands and in lymphatic and hepatic metastases. On the other hand, osteosclerosis does not seem to be observed and bone metastases are rare.

The most striking clinical symptom that is really quite embarassing to a minority of the patients is the diarrhea. The percentage of sufferers from diarrhea in a population with widespread MTC has been said to be around 30. However, this seems much too high. In some patients the diarrhea has disappeared after radical surgery, and there is no question that the diarrhea in MTC must be produced by some substance from the tumor. Diarrhea has been described in some patients with neuroblastoma. The connection between neuroblastoma and pheochromocytoma is very close, but it is clear that the diarrhea also occurs in patients, who have no associated pheochromocytoma.

Much work has been done regarding the correlation between diarrhea and the formation of active substances in these tumors. A number of polypeptides, such as substance P, active in promoting peristalsis have been investigated recently. Calcitonin is certainly not the explanation since we have seen patients with high calcitonin levels who have had no diarrhea, and the physiological effect of this hormone on the gut does not seem to be remarkable. I have already mentioned the parallel between MTC and carcinoid tumors. Many authors have looked for serotonin as a possible cause of the diarrhea in MTC but have found no positive evidence. Urinary concentration of 5-HIAA has been determined for many patients but it has not been much increased.

A recent paper by Bernier et al treats the problem of diarrhea (41). These authors tried to confirm the findings of Williams, Karim, and Sandler that prostaglandins might be the cause of the diarrhea (493). However, they could not find any indications of increased prostaglandin levels. It must be remembered, however, that these substances may be metabolized rapidly, and it is possible that analysis of such breakdown products may be rewarding. Among the other active substances some of the intestinal polypeptides such as VIP, motilin, and substance P (176) may be a possible explanation. This problem is discussed in this Chapter, Section 4.

Many of these tumors seem to produce several substances, and it is possible that mechanisms for the production of one special active substance may be activated in the cells of one type of tumor and be inactive in another in which some other substances are produced. As far as we know at present calcitonin always is produced in large quantities in tumors with the typical histology of MTC. This is obviously a topic product just as the catecholamines are produced by pheochromocytomas. Production of ACTH causing the clinical picture of Cushing's syndrome is perhaps the most well-known ectopic paraneoplastic condition in MTC. This problem has been discussed previously, but it is a fact that only a very small number of patients with MTC have had a typical hyperadrenocorticism. We have never seen it. E. D. Williams has collected the literature on this problem (494) and points out that this combination is rare. He found it in only 3% of his patients. It is also stressed that MTC is the only type of thyroid tumor with ACTH production (95). On the whole classical thyroid carcinoma is hardly ever involved in paraneöplastic processes. No patient seems to have recovered from Cushing's syndrome after thyroidectomy, but ACTH has been demonstrated in several MTC tumors and that connection is now well established.

There has been much discussion on the true nature and origin of the cells that produce calcitonin. It seems evident that these so-called C-cells contain this hormone since they give a strong fluorescence when treated with fluorescent antisera to calcitonin. It has also been proved that the ultimobranchial body in the chicken contains a high level of calcitonin and must be the source of this hormone. It does not fuse with the thyroid gland in this species. Thus the question regarding the presence of C-cells in the thyroid gland in birds is important. A very ingenious experiment with a technique used in developmental biology ("Entwicklungsmechanik") has been performed by Le Douarier and Le Lièvre in 1970. They seem to have proved that some cells in the ultimobranchial body of this bird, with ultrastructural characteristics indicating secretion, come from the neural crest. Cells from the quail may be recognized histologically. Small pieces from early embryonic spinal cord of the quail were introduced into chick embryos. When these embryos developed further, quail cells were demonstrated in the ultimobranchial body. This seems to be convincing evidence of some cell migration from neural crest into this structure, but it hardly proves that these cells are secreting calcitonin since we know that the large body of cells in the ultimobranchial body comes from the last branchial pouch (see also 263).

3. MASTOCYTOMA (MASTOCYTOSIS)

Mastocytosis (urticaria pigmentosa) has been known as a disease of the skin since the 1870s. The lesions are mostly papulomacular. Their colour is red or sometimes more yellowish. When the lesions heal, they may leave pigmented spots. In certain cases these lesions are rather isolated, in others they are crowded, giving the skin a very mottled appearance. The characteristic feature is the strong urticarial response obtained on stroking the papule. This is obviously caused by the liberation of histamine in the skin. The infiltrates consist of tissue mast cells. In 1949 Ellis demonstrated that one of his patients with the typical skin disease also had the same infiltrates in the internal organs (121). Since that time a large number of authors have published their observations on such patients (17, 108), who should be regarded as suffering from generalized mast cell disease (mastocytosis) (310). It is interesting that osteosclerotic lesions containing mast cells are not rare in the axial bones (66).

The mast cell is the fundamental constituent of the pathological tissue. They were first described by Paul Ehrlich, who demonstrated their special affinity for basophil dyes giving a so-called metachromatic color. He called them "Gewebsmastzellen" and distinguished them from the circulating basophils in the blood, "Blutmastzellen." These tissue cells are normally located in many organs, and they seem to be abundant around the vessels. From a functional point of view, they constitute a special organ, and their metabolic processes have been studied extensively during the last decades (for literature and recent reviews see 108).

The metabolism of histamine has been worked out in great detail, and it is clear that a number of metabolites are excreted. Their importance for the diagnosis of mastocytosis and for the control of different therapeutic procedures is not thoroughly known. It is clear, however, that imidazole acetic acid is one of the most important breakdown products of histamine.

Infants and young children constitute a large group among the patients with urticaria pigmentosa. This type with often localized mastocytosis may sometimes disappear later in life, but it usually lasts a number of years. The adult form of the disease is nearly always progressive. Some patients develop an enormous number of small papules and also osseous infiltrates. On the other hand, we may see patients with very small cutaneous tumors and widespread osteosclerosis. Thus, there is no certain relationship between the two.

Usually there are very numerous small discrete infiltrates on the trunk. Sometimes these infiltrates coalesce, and it is common to see pigmented patches. Both purpura and acne have been noted as incorrect diagnoses. In some patients the skin shows hyperemic patches, in others, slight or bright transitory flushes may occur. This seems to be more common in children. We have seen one patient who had very widespread cutaneous infiltrates and developed quite severe flushes together with circulatory symptoms. The severest reaction, that we have seen, occurred in the patient described below. Her case history was first published by Hamrin (181).

The patient, a woman born in 1930, had had lesions of the skin for about 7 years, when we saw her in 1957. These lesions had not given her any trouble. The results of a biopsy were typical. Her resting blood pressure was 130/80 mm Hg. Liver and spleen were not palpable. Radiological examination of the skeleton showed marked diffuse osteosclerosis in the thoracolumbar spine, in the pelvis, in the skull, and in some bones in the extremities. Sternal puncture showed the findings at biopsy were typical.

Her hematological status was normal. The urine contained 210 micrograms of free histamine (6 to 19 times normal) and a low level of HIAA. The patient was very much worried by the fact that she thought she had a hypersensitivity to aspirin. She reported that she had once reacted to this substance with intense vomiting and fainting. She wanted us to confirm her hypersensitivity. We therefore gave her a provocation dose as described in the paper by Hamrin. Before this she was given half a gram of the constituents contained in aspirin tablets. This did not produce any symptoms. She then took 0.25 g acetylsalicylic acid perorally. For the first 20 minutes no reaction was noticed, but then she suddenly felt nauseated and vomited. A red flush started on her face and spread rapidly downward over neck, chest, and arms. During the next 30 minutes the rash became generalized and she showed edema on her face and in patches on her arms. She also experienced a tingling sensation all over her body. It is seen from Figure 14 that she developed very marked tachycardia and a drop in blood pressure to systolic 40 mm Hg with a corresponding rectal temperature of 36° C. After 3 hours the acute response subsided but she continued to vomit during the next 14 hours. We measured the volume of gastric juice that she lost during this last period. It was more than 2 liters of strongly acid fluid. During the next 24 hours her pulse rate stayed high and she had a fever that disappeared slowly.

The excretion of histamine in the urine as well as the levels of HIAA were followed. The maximal values were found as late as 14 hours after the start of the attack. The excretion then rapidly returned to the initial level. The patient had no signs of asthma even though she felt some pressure in the chest. She had a severe headache for about 12 hours. The HIAA excretion rose from an initial value of 80 μg/hour to a highest level of 285 μg/hour, when there was also a maximum excretion of histamine. After the attack the patient rapidly recovered and everybody was certain about the fact that she was hypersensitive to aspirin!

This is obviously not an allergic hypersensitivity reaction but more like the action of a histamine liberator in a patient with very massive stores of histamine in her mast cell organ.

It has already been mentioned that Ellis and later a number of other authors have found signs of widespread infiltrates in many organs. Enlargement of the liver and spleen has been observed even though this is not a common finding. More important is the fact that the skeletal system may show rather specific changes. The most difficult symptom to interpret is the osteosclerosis, sometimes patchy, sometimes of a more diffuse spread through many bones but chiefly in the pelvis and the lower vertebrae. The diagnosis of mastocytosis always should be remembered when a patient with osteosclerosis of unknown origin is seen. The lesions are not always sclerotic; sometimes there may also be found signs of lysis, and then the interpretation is still more difficult.

Sternal puncture may give very valuable information for the diagnosis in some cases. We have followed one man who had had mastocytosis for many years.

The patient was 76 years old when he died in 1976. He had suffered from severe cardiovascular attacks for many years with marked flushing and hypotension probably caused by release of histamine. A bone marrow biopsy specimen showed numerous atypical basophil cells. We were almost certain that he would slowly develop a

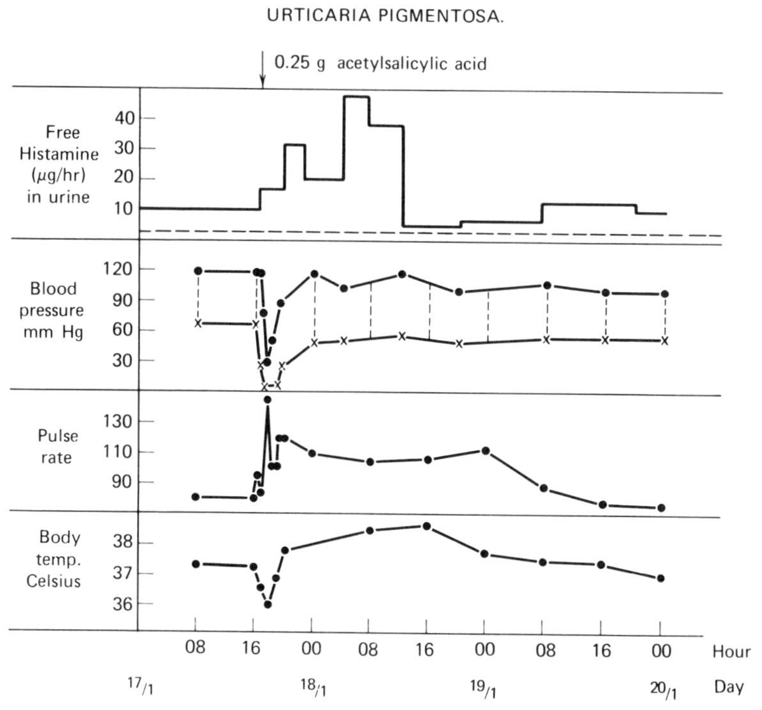

Figure 14. Shock provoked by 0.25 g acetylsalicylic acid. The broken line in the top part of the figure indicates the upper limit of normal histamine excretion. The shock is described in detail in the text. (From Hamrin in Brogren et al [66] with permission of the authors and of *Acta Med. Scand.*)

hematological condition of the type we call mast-cell leukemia. (This disease is obviously very rare. Only a few instances have been described where large numbers of tissue mast cells entered the circulation and were found in the peripheral blood. Flushing is then common. The presence of this clinical picture proves that mastocytosis may become a malignant systemic disease comparable to leukemia and lymphoma.) Our patient died of an acute cardiac failure possibly caused by ingestion of salicylic acid. He had no acute infarct. Large numbers of mast cells were found in most organs.

It now seems quite clear that the symptoms are caused by histamine produced in the mast cells (181). There are some observations on patients, who have had localized rather large mastocytomas. Under pressure these tumors pour out histamine into the circulation causing flushes. Mechanical stimulation by pressure, when performing tests for dermographism, also causes edema over the cutaneous infiltrates. We do not know of other physiological mechanisms for the liberation of histamine but it is probable that a number of the well-known histamine liberators that cause degranulation of mast cells are effective in mastocytoma. We have never dared to repeat the experiment quoted.

It is known that heparin is stored in large quantities in the mast cells. Many investigators have tried to find signs of hyperheparinemia in patients with mastocytosis. So far this has not been possible. Even very sophisticated analysis of the coagulation pattern has not indicated any increase in circulating active heparin. It is possible that heparin is more firmly bound to some basic protein, but it is also evident that there must be a physiological mechanism for its release from the cell. Evidently much is to be done in this field. The fact that dogs not uncommonly carry mastocytomas offers an experimental approach to these problems.

Only few patients with mastocytosis have died, but the postmortem examinations have shown that the osteosclerotic parts of the bone are infiltrated with mast cells. The connection between these phenomena is obscure, but it is interesting to speculate on the fact that carcinoid bone metastases in patients with gastric tumors producing histamine are also osteosclerotic (476). At present it is impossible to decide if histamine is in some way responsible for this disturbance of calcification. The only normal substance that is known to promote calcification of this kind is calcitonin. In our elderly patient with severe flushes and osteosclerosis calcitonin was determined in the serum and was found to be normal.

The treatment of the disease is not very effective and the most important rule seems to be to avoid histamine liberators. Preparations containing salicylic acid may be potentially dangerous.

4. PHEOCHROMOCYTOMA, NEUROBLASTOMA, AND GANGLIONEUROMA

Tumors from the autonomous nervous system all come from the neural crest. Cells may develop only into the blast stage (neuroblastoma). Further differentiation then leads to ganglioneuroblastoma and finally at the most mature stage to ganglioneuroma. The blastoma stage is extremely malignant and metastasizes in the bone marrow giving a cytological picture that may resemble blast-cell leukemia. For the differential diagnosis the presence of secretory granules (even in small cells) is decisive. The presence of mature ganglion cells is a good omen. Other metastases, which may calcify, are found in lymph glands and the liver. Cells from the paraganglia and the adrenal medulla develop into pheochromocytoma, but it must be remembered that the blastomas may occur inside the medulla of the adrenals even though the majority of these tumors are retroperitoneal (two of three) or occur in the thorax.

It is probably true that pheochromocytoma (PHC) is the first tumor with a definite diagnosis built on the discovery of a specific active metabolite in the urine. In 1950 Engel and von Euler described a patient with the tentative clinical diagnosis of PHC and a massive excretion of epinephrine in the urine (124). The diagnosis was made with the help of bioassay. Surgery was successful. During the last decades a large number of chemically identified metabolites have been found, and these are all related to catecholamine production. I shall not discuss the clinical picture in PHC as it is well known. It should be pointed out, however, that hypertension is not always the most striking symptom. The catecholamines are calorigenic and an increased metabolic rate with persistent tachycardia may lead to weight loss. Sweating is almost a constant but not quite compulsory symptom. Paradoxically patients have been described who have had attacks of hypotension that brought them to the doctor. I saw one such patient, who was later published by Hamrin (182). The general rule at present is to have determinations of catecholamines and metabolites performed on all patients with severe unexplained hypertension.

Malignant PHC is not common but has been described especially in patients who have had bilateral tumors. Many of these tumors obviously have a genetic origin, and families with a combination of multiple PHC and medullary thyroid carcinoma are known (Sipple's syndrome). In these patients it is important to follow the serum and urinary patterns of different metabolites after surgery to be certain that the operation has been radical. For diagnosis and follow-up, determination of different metabolites is most convenient, but in some patients, it may be necessary also to determine the level of free catecholamines in the urine.

For the diagnosis of neuroblastoma (NBL), the investigation of urinary metabolites is very important. This tumor is one of the most interesting in oncology (226, 227). It is highly malignant, and has been found at autopsy both in the fetus and in newborn infants. Infants at the time of diagnosis often are cured, whereas older children almost always die. The peak incidence is earlier than 3 years. No less than 30 cases of NBL were observed to have "spontaneous" regression. In four of these the NBL was transformed into a mature ganglioneuroma. One of these patients received no treatment at all. The child had multiple subcutaneous tumors that were excised at different stages. Finally, they all developed into ganglioneuroma. At age 14 the patient had calcified tumors in the liver and in the left adrenal and retroperitoneal glands. It is obvious that a number of patients have been cured even when inadequately treated, and that the regression therefore must have been spontaneous. The immunology of these tumors has been studied very carefully. Hellström and Hellström have reported on the immunological findings in such cases, and this tumor has been a paradigm of spontaneous regression, possibly accompanied by immunological changes (192). Bill (43) investigated 200 infants and fetuses, who had died of external well-defined causes, for the presence of NBL. He found that 0.4% had NBL. Luckily the clinical signs of the tumor are very rare among newborns and infants (1 of 10,000). It seems possible to assume that a normal maturation factor is present for these cells that may develop late in some newborn infants. The parallels between other factors in renal developmental ontogeny is evident. I feel certain that the problems of Entwicklungsmechanik will become most important for oncology (171).

I have mentioned these biological characteristics of the tumor to illustrate the importance of biochemical investigation of the urine (226, 227). Ninety-five percent of the children with neuroblastoma showed increased catecholamine and/or metabolite levels that returned to normal rapidly after radical surgery, and more slowly after radiation or chemotherapy. If the levels remain high this indicates that the treatment has not been radical. The excretion patterns may be very different, and it is incorrect to say that the determination of VMA alone is sufficient to exclude neuroblastoma, even though it is rare to find normal values. The epinephrine level is always normal because the tumor cells do not have the metabolic enzymes necessary for its synthesis. It is important to know that the excretion of all these substances increases with increasing age until puberty. Pure neuroblastoma cells usually form less norepinephrine than the cells of ganglioneuroblastoma and PHC, but even when they excrete large amounts of norepinephrine, there are rarely any clinical signs of hypertension. This may be connected with the fact that these tumors have few storage granules as compared with PHC. It is impossible to make a biochemical differential diagnosis between neuroblastoma and ganglioneuroblastoma (see 423–425).

The third tumor of this type in children, namely ganglioneuroma, only shows an increased formation of active substances (except in the urine) when they have the syndrome of intractable diarrhea (171). An interesting report of a case of this kind was published by Rosenstein and Engelman from Johns Hopkins Hospital in 1963 (366). This paper contains an excellent tabulation of tumor cases.

Their patient was a 15-month-old boy who had had diarrhea since the age of 6 months. He was chronically malnourished with thin, wasted extremities and distended abdomen and had intermittent flushing over his entire body. This had been noted by his mother for 3 months and was independent of the other symptoms. His stool in 24 hours was 425 ml and contained 45 mEq of potassium. However, his serum potassium was still 3.4 mEq/100 ml. Abdominal X-ray films showed calcification in the left suprarenal area, and later, indications of a lemon-sized tumor above the left kidney were found. The tumor was extirpated. The flushes disappeared, and by the seventh postoperative day, he had only one or two normal stools per day. Nine months later he showed no signs of relapse, and his urinary level of 5-HIAA was normal. The tumor contained increased amounts of metabolites, namely, catecholamines and dopamine.

The localization of these tumors may be surprising. Chamberlain has published his observations of an interesting case (79).

The man, aged 54, had had a lump in the neck that increased in size. He suffered from cramping abdominal pain and severe diarrhea but no bleedings. On admission a florid complexion was noted. His liver was slightly enlarged. His urine contained no HIAA nor catecholamines and his glucose tolerance was normal. The tumor was removed in two stages. The diarrhea and pain improved after the operation and he had only two bowel movements a day instead of seven or eight. There was no detailed biochemical or pharmacological examination (75).

Stickler has reported on a case from the Mayo Clinic (415).

A 19-month-old boy suffered from very severe diarrhea. After treatment with prednisone there was some improvement, but his abdomen was still very distended. Abdominal X-ray films showed a picture of malabsorption. His 5-HIAA level was normal, and his serum sodium level was decreased to 105 mEq/L and potassium level to 1.4 Meq/L. The abdominal tumor was very large and inoperable. A biopsy specimen showed a ganglioneuroblastoma. Excretion of catecholamines and metoxymandelic acid was increased, but the authors regard these substances as irrelevant to the diarrhea.

A study of 22 patients with ganglioneuroma and five with ganglioneuroblastoma in the literature showed that only one patient had severe diarrhea. This therefore must be a rare symptom in this condition. Regarding the biochemical picture in the urine the reader should consult papers by von Studnitz and by Käser (226, 227, 424). Urinary metabolites were examined in 25 patients with neuroblastoma. HVA was increased in 17 and mandelic acid, in 24. Five other cases with benign neurological tumors had normal levels. The authors were not able to relate different metabolic pictures to a different morphology (226, 227, 424).

11
Ectopic Production of Hormonally Active Polypeptides

1. THE HYPERCALCEMIC SYNDROME

Astute observers have long been aware of the fact that calcifications may occur in metastatic mammary carcinoma. Virchow published a report of a case. It is impossible, however, to tell if his patient also had hypercalcemia, but we know from studies of myeloma that tissue calcinosis might occur when plasma levels of calcium and phosphate are high. On the other hand, calcification of hepatic and lymph gland metastases may occur together with disturbances of calcitonin production, e.g., in medullary carcinoma of the thyroid. It is therefore impossible to tell if such early observations, without chemical analysis of the serum, were caused by general hypercalcemia or by local hormone effects.

In the introduction to this book, I mentioned that Fuller Albright observed a patient with a renal tumor who had marked hypercalcemia without much evidence of bone metastases (3). He hypothesized that the tumor tissue might produce some factor that increases the calcium content of the blood, and he even sent a piece of the tumor to Collip, who was at that time the great authority on parathyroid hormone. No such hormone could be demonstrated.

In 1956, Plimpton and Gellhorn published a report of 10 cases with cancer and hypercalcemia (339). The most striking observation was of a woman who was suffering from very severe hypercalcemia. She also had an ovarian carcinoma, but it was thought necessary to alleviate the hypercalcemia as the first stage in treatment. An exploratory surgery of the neck was performed, but no parathyroid tumor was found. After extirpation of the ovarian carcinoma, her serum calcium levels rapidly returned to normal and remained so until metastases occurred, and the patient died hypercalcemic. This is an excellent example of reversibility after tumor extirpation.

During the next 20 years there was such a wealth of publications on the subject that it is impossible to cite them all. The reader is referred to an excellent paper by Lafferty written in 1966, having the rather obscure title "Pseudohyperparathyroidism," in which he discusses 50 case reports from the literature (244). The cases were selected according to three criteria, of which at least one was noted: (a) no bone metastases found roentgenographically or at postmortem examination and no parathyroid adenoma found at exploration or at careful autopsy, (b) significant reduction of the serum calcium level after resection of the neoplasm, and (c) positive immunoassay for parathormone (PTH) in an extract from the tumor. Of the 50 cases, I feel that those in which there was normalization after the operation are the most interesting. These consist of 10 cases of hypernephroma, one of carcinoma of the renal pelvis, and one of the bladder, three with bronchogenic carcinomas, two with ovarian carcinomas, one of myosarcoma of the uterus, one of squamous cell carcinoma of the vulva, and one of carcinoma of the penis. It is clear that there is much variation among the tumors, but some remarkable exceptions should be noted. Most striking is perhaps the fact that there is no mammary carcinoma among these cases. In the whole group of 50 cases, there is no case of mammary carcinoma nor of prostatic carcinoma. It has been thought to be a sign that these tumor cells do not produce any PTH-like factor. This may be true, but another explanation could be that these tumors metastasize to the skeleton very early and therefore are excluded from most studies because of the criterion that there should be no bone involvement (see also 13, 31, 35, 45, 70, 88, 169, 288).

Most important for this discussion is the question of whether or not the presence of bone metastases really has anything to do with the hypercalcemia. During periods, when the morphological outlook was dominating, it was axiomatic that bone minerals may be liberated when neoplastic cells grow in the skeleton. From my own experience with myeloma, I have become rather skeptical about this explanation. Every physician has seen widespread severely destructive myeloma that was never accompanied by any increase in serum calcium levels, whereas other patients with very little visible bone involvement may be severely hypercalcemic. I have always stressed the point that the presence of Bence Jones protein is strongly connected with hypercalcemia. To me it seems quite possible that a part of some individual Bence Jones molecules may have an amino acid sequence that influences calcium metabolism. There are also many arguments in favor of the idea that kidney function is deranged in many ways by Bence Jones proteins (398) (see Chapter 5).

In a recently published survey of hypercalcemia in malignant disease, the group from the Mayo Clinic has attacked those problems from another point of view (35). They point out that "ectopic hyperparathyroidism" is about as common as the primary, and it is difficult to locate the tumors in very many patients. One hundred and eight unselected hypercalcemic tumor patients were investigated with a sensitive radioimmunoassay and antisera specific for the carboxy terminal part of the PTH molecule. One hundred and three patients had abnormally high PTH concentrations. Forty-eight of these had evident bone metastases, *and their hormone values were nearly as high as those without such metastases*. This could mean that the hypercalcemia is more related to the formation of a humoral factor than to the "pressure" effect of tumor tissue on bone. Of the patients with high serum calcium levels, most had hypernephromas. Then in order of frequency were squamous cell carcinoma of the lung, different lymphomas, mammary carcinoma, and myeloma. All the myeloma cases naturally had some bone marrow involvement. A large number of other localizations were found, but it is interesting that adenocarcinoma of the lung as well as in the colon and in the ovary were rare, 1 and 2% respectively. (The colonic and ovarian adenocarcinomas had not metastasized to the bone.)

One patient group is of special interest. It has been stressed by Gordan that the hypercalcemia commonly seen in paitients with metastasizing carcinoma of the breast should be caused by some factor that is not PTH-like. He points out that breast cancer is the most common malignancy causing hypercalcemia, and it has been said that two-thirds of all patients with paraneoplastic hypercalcemia have breast cancer. In Gordan's laboratory, Gardner was able to prove in 1963 that breast cancer tissue and lung cancer tissue produce lysis if cultivated in contact with a skull from the rat, whereas normal breast tissue is inactive (150). There are cases of mammary carcinoma with high serum calcium levels and no metastases even at autopsy. It is possible that we need a new method to find out if a certain group among the patients with mammary carcinoma have a factor other than a PTH-like one which explains their disturbed calcium homeostasis. It is important to remember that certain patients develop hypercalcemia when treated with hormones (estrogens or androgens) (See 88 and 230).

Gordan has pointed out that patients with mammary carcinoma do not have low serum phosphorus levels, even when they are hypercalcemic. This is used as an argument against the presence of a factor with PTH-like activity. The group at the Mayo Clinic has studied this special problem and their results were as follows (35). When comparing serum calcium values in primary hyperparathyroidism (87 patients) with the values found in cancer, the latter had the highest levels. Hypercalcemia was most severe in mammary carcinoma. One very important factor in the regulation of serum phosphorus is the kidney. It is an everyday experience that a patient who develops a myeloma kidney and hypercalcemia shows lower calcium values when kidney function is more severely deranged since this results in phosphate retention. A numerical improvement in the calcium level really means that the patient's condition is deteriorating. The major patient group in the study consisted of 68 azotemic patients of 108 tumor patients, whereas only 26 of 87 hyperparathyroid patients had azotemia. When these azotemic patients were excluded, all the mean serum phosphate levels were significantly lower than normal. The authors argue that this indicates a similarity between paraneoplastic patients and those suffering from parathyroid hyperfunction. This seems to be a convincing argument, and it shows how important it is in clinical medicine to analyze as many homeostatic factors as possible.

From the practical point of view, it is most important to find out if patients with a high calcium level caused by malignant tumors can be differentiated from those with hyperparathyroidism. The group at the Mayo Clinic stressed that the value for immunoreactive PTH is relatively lower as compared with the serum calcium values in the cancer group. This may be a valid finding in a laboratory equipped with very special antisera, but the authors also point out that primary hyperparathyroidism produces biologically inactive hormone fragments that are detectable immunologically. If these ideas are confirmed, they will show how complicated these problems have become and only continued critical research in a few very well-equipped laboratories will help to solve the problems definitely.

The question of pro-PTH is also interesting from the paraneoplastic point of view. It has been shown that this hormone has an aminoterminal hexapeptide placed before the alanine in ordinary PTH. This difference is probably true for both bovine and human hormone. The pro-hormone may be excreted directly from the cells, but it is also possible that it is broken down inside the cell and excreted as PTH.

From a large pool of bovine glands, a number of polypeptides with the same molecular size and charge as the PTH have been prepared. A number of these are not active, but there seems to be three different forms of PTH that are active. They all have 84 amino acids, a terminal alanine, and no cystine (232). The dominating form has been sequenced completely. Bovine PTH is active in man.

A. Hypercalcemia Caused by Factors Other than Ectopic PTH

In 1971 Roof and others from Gordan's group found that the ectopic "parathyroid hormone" did not behave like the normal hormone on immunological analysis (361). These authors believe that there must be two or more types of ectopic parathyroid hormone, one that cannot be distinguished from the normal type and another that is different immunologically. Patients with different tumors synthesize different types. It is rather interesting that the Gordan group has found normal levels of PTH in hypercalcemic patients with breast cancer, and that the paper from 1971 only treats the conditions in other patients.

The effect of treatment with glucocorticoids is said to be much better in patients with normal PTH levels and hypercalcemia. Mannheimer studied the effect of corticosteroids in a large group of patients with mammary carcinoma. The initial dosage was quite large but a maintenance dose of 15–20 mg of prednisone daily was sufficient to keep the values at normal levels. Primary hyperparathyroidism was never found at the postmortem examination but metastatic calcifications were common. Favorable response was seen in 32 of the 40 patients, and 46 of 59 episodes of spontaneous hypercalcemia were treated successfully (282). Another study showed the same incidence of therapeutic effects. In this study, all patients had widespread osteolytic metastases.

A paper from Massachusetts General Hospital by Powell and his group gives an analysis of 11 cases with hypercalcemia and hypophosphatemia in cancer patients without bone metastases (341). In nine patients surgery or antitumour therapy produced a biochemical remission. Parathyroid hormone was assayed by radioimmunological methods, and three different antisera were used that were active against practically all regions of the PTH molecule. Still more important seems to me the result of bioassay, when tumor tissue extracts were tested for osteolytic activity on the mouse calvarium. The patients had variable types of tumors, and no patient had typical mammary carcinoma. Incubation of bone tissue with extracts from normal liver tissue or from cancerous tissue from a patient without hypercalcemia gave no bone resorption, whereas extract from parathyroid tissue or from a lung carcinoma from a patient with hypercalcemia—but no detectable PTH—was strongly positive. These results may be difficult to explain, and it seems probable that only more refined biological tests will solve the problem. One way might be to investigate the biological effect of different antisera on normal PTH and on tumor extracts when tested on bone tissue.

The only neoplasm for which hypercalcemia has been considered as part of the clinical picture is multiple myeloma. (It was first discussed by Gutman in 1936 [174].) In this disease the relationship between nephrogenic hyperphosphatemia and serum calcium level is also very important. I have seen a number of patients in the final intractable azotemic stage of the disease who have had continued increasing hypercalcemia and then show spontaneous decrease in calcium levels (see Figure 15). This is not a sign of improvement if the phosphatemia has increased at the same time.

B. Prostaglandins and Hypercalcemia

In 1965 Wilson et al published a paper on a hyperparathyroidlike state in rabbits with VX_2 carcinoma. This tumor produces severe hypercalcemia 3–4 weeks after

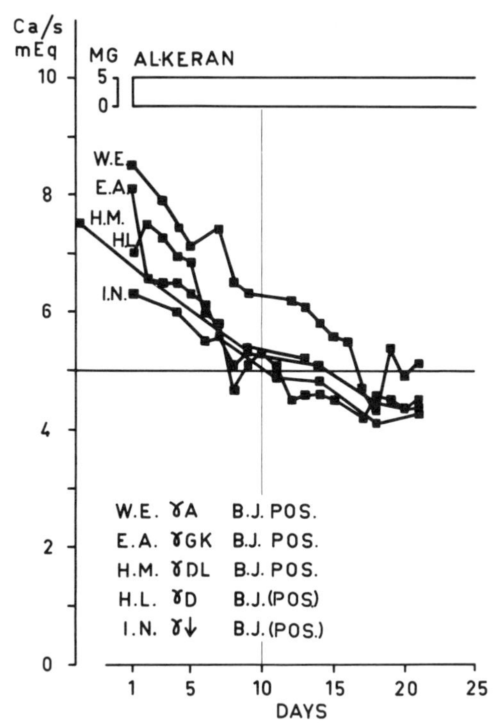

Figure 15. Serum calcium levels in five patients with myeloma and with hypercalcemia treated only with Alkeran. Combination with glucocorticoids probably gives more rapid response. Hydration is most important. In the future treatment with calcitonin may become the treatment of choice.

transplantation without the appearance of bone metastases. It is reversed by operation and does not depend on the parathyroid glands (498).

In 1970 Klein and Raisz demonstrated that prostaglandins (P.G.) stimulated bone resorption in tissue culture (236). Auerbach and Chase tried peptide hormones other than PTH to determine their effect on the concentration of cAMP in fetal rat bone. They also tried P.G. and found that it increased the concentration of cAMP. The similarity between PTH and P.G. led them to conclude that P.G. might promote bone resorption. This was investigated in tissue culture, and it was found that in fetal bones containing previously incorporated ^{45}Ca the isotope was mobilized by P.G.

The tumor cells secrete large quantities of P.G. E_2, and inhibitors of P.G. synthesis normalize the serum calcium values. Tashjian and his coworkers noted a bone resorptive factor produced by certain mouse fibrosarcoma cells. They investigated the problem of hypercalcemia and tumor P.G. using the rabbit carcinoma as a model (431). There was no question about the fact that the tumor cells produced a bone resorption factor. With serological methods it was found that the tumors contained P.G. E_2 and that this was a potent bone resorption stimulating agent. Indomethacin—an inhibitor of P.G. synthesis—inhibited these effects even prophylactically. However, after cessation of therapy hypercalcemia appeared but could be reversed again. The results seem to be most convincing (see also 185).

In 1974 Brereton et al published an account of indomethacin-responsive hypercalcemia in a patient with renal carcinoma (64).

> Their patient, a 54-year-old man, had multiple liver metastases after resection of a renal cell adenocarcinoma. He had persistent hypercalcemia but no indications of bone metastases. Tests for serum PTH were negative (Roof). Neither hydration nor phosphates had any effect, and mithramycin was only slightly effective. His serum calcium levels ranged from 5.6 mEq/L to 6.7 mEq/L. Indomethacin, 25 mg b.i.d., was started, and his serum calcium level became normal. When indomethacin was stopped, there was a rapid relapse with good effect from the next course. The patient died of hepatic failure. At autopsy no bone metastases were found and three parathyroid glands were normal. The liver metastases contained large amounts of P.G. The lung metastases contained less than the normal lung, which is somewhat surprising.

The next observations of the therapeutic effects of indomethacin on hypercalcemia were published by Seyberth et al (388). These authors studied the excretion of a metabolite of the E-prostaglandins (P.G.-E-M) in the urine. It was significantly greater than normal in cancer patients with hypercalcemia but not in cancer patients without hypercalcemia. Only six patients with hypercalcemia from surgically proved hyperparathyroidism were investigated. They all had normal P.G. E-M values, whereas their immunoreactive PTH was high. All the cancer patients with hypercalcemia had normal PTH values. Fourteen hypercalcemic patients with tumors were investigated: 10 had carcinoma of the lung (nine squamous cell), two had adenocarcinoma of the pancreas and two others had metastatic undefined adenocarcinomas. Three patients with myeloma, one with lymphosarcoma, one with reticulum cell sarcoma, and one with lymphoma had hypercalcemia with normal P.G. E-M values and normal PTH values. This is surprising.

Four hypercalcemic patients with cancer were treated with aspirin and two with indomethacin for 5–7 days. No other therapy was given. In 5 of 6 patients P.G.E. metabolism returned to normal (probably also in the sixth). In two patients the serum calcium level remained the same. Both patients had squamous cell carcinoma of the lung with bone metastases. No patient with myeloma was treated since all three had normal P.G.-E-M levels.

The data are not so clear-cut since it is very difficult to determine the P.G. metabolites.

C. Other Factors

Mundy has made some interesting studies on the presence of a bone resorbing factor in the supernatant from cultures of lymphoid cells and of myeloma cells (306). Short-time cultures of bone marrow aspirated from seven patients with myeloma were studied. From the fluid in the tissue culture a factor was obtained that stimulated the liberation of previously incorporated radioactive calcium from organ culture of bone. Bone marrow cells from seven other patients with different blood diseases did not produce any stimulator. It is suggested that stimulation of osteoclastic activity is the reason for bone destruction in myeloma.

There is no doubt about the fact that hypercalcemia is one of the most important paraneoplastic syndromes for two reasons: (a) Its occurrence in patients with malignant tumors is likely to be more common than in primary hyperparathyroidism; (b) active treatment is very effective and abolishes a situation that is life-threatening.

The present tendency among active internists is to have *determinations of serum calcium as a routine test* and it is unquestionable that such biochemical screening should include serum calcium. A large number of investigations have been performed during the last years and the results are quite uniform. Such screening may detect a number of conditions with uncharacteristic symptoms and lead to a diagnosis that is amenable to treatment. In a Swedish country hospital systematic screening of a large number of patients regarding the presence of hypercalcemia was performed (414).

The results of some surveys of the frequency of paraneoplastic hypercalcemia in populations of unselected patients in different parts of the world have been published. McLellan et al analyzed 3,700 determinations of serum calcium levels in suspect patients in Australia (276). Values above 10.4 mg/100 ml were regarded as elevated, and 151 persons (0.3%) had such levels. The true cause could be found in only 61 cases. Half of these patients had tumors. Breast cancer was leading type (twelve), followed by lung cancer (five), myeloma (four), cancer of the kidney (two) and rectum (two), and various other types. There were only seven cases of hyperparathyroidism from adenoma and one case of parathyroid carcinoma. Six patients had sarcoidosis. Of the 12 patients with breast cancer, (five) had no bone metastases.

D. Clinical Symptoms

Most authors seem to agree that the clinical picture in paraneoplastic and in primary hypercalcemia caused by hypersecretion of normal PTH is very similar.

The number of cancer patients who have undergone exploratory neck surgery seems to prove this. High age may of course be an argument in favor of cancer, but it is of no real value in individual cases. The paraneoplastic history is usually very short, whereas a large number of patients with primary hyperparathyroidism had a long history of renal stones and also of abdominal symptoms. Weight loss is, as always, an argument in favor of cancer. Thirst and polyuria are common, followed by dyspeptic symptoms and in many cases disturbances of the CNS—first, lassitude and loss of memory, and later mental confusion and finally coma. Many patients with myeloma, who have been regarded as having paraproteinemic or azotemic stupor, and later coma, in reality suffered from hypercalcemia and were much improved when this symptom was controlled. Disturbances of the electroencephalogram are common and reversible, but it is difficult to distinguish between symptoms caused by uremia and hypercalcemia since these two conditions improve together (130).

Regarding the laboratory values it is said that an increased alkaline phosphatase level is more common in paraneoplasia, but the presence of liver metastases usually troubles the picture. It is hard to differentiate biochemically between bone and liver alkaline phosphatases. Many authors agree with Wills and Gowan (497) that serum chloride is an important marker for the distinction between paraneoplastic and primary hypercalcemia. It seems clearly established that chloride values below 102 mEq/L were common in hypercalcemia that was not caused by primary hyperparathyroidism. Several authors have found that very low values are much more common in tumor patients, but the influence of vomiting on these parameters must be considered. However, the serum chloride level does not seem to be of much use for differential diagnosis. In our experience with a large number of patients with myeloma, we have found that the many factors influencing serum chloride levels make diagnosis based on this parameter very difficult.

E. Treatment of Hypercalcemia in Malignancy

Antitumor Agents and Corticoids

It is clear that extirpation of the primary tumor is the only radical way of treating patients with paraneoplastic hypercalcemia, and the results at times have been excellent and convincing as was demonstrated already by Plimpton and Gellhorn in a patient with ovarian carcinoma. Recent investigations have shown that treatment with chemical antitumour agents also may be very effective. The present author has published a number of graphs showing that serum calcium levels in patients with myeloma who were treated only with large doses of alkylating agents (Alkeran) declined rapidly from hypercalcemic to normal values (466). The addition of steroids did not seem to improve the response. It is clear, however, that the time necessary for normalization is about 10 days, and this may be too long for a patient with severe hypercalcemia. Even if treatment with appropriate antitumour drugs is the most important measure in patients with widespread malignant disease such as myeloma and metastatic cancer, more rapid methods must be found.

The experiences with cortisonelike hormones in the paraneoplastic conditions are somewhat variable. Cortisone alone is usually quite effective in the treatment of myeloma. A combination of cortisone and some alkylating agent is perhaps still more effective, but it should not be forgotten that the latter preparation may have a rapid effect when given alone. Almost all patients with myeloma with high serum calcium levels are dehydrated, and systematic hydration in itself may lead to great improvement. It is therefore difficult to judge the influence of different drugs when hydration is maintained—as it should be. Regarding carcinoma, it has been said that different tumours react differently to treatment, and it is possible that patients with bronchogenic carcinoma as well as those with renal and ovarian cancers are slow responders. In many patients with cancer, hypercalcemia is a very late sign and therefore not very important from the point of view of treatment. On the other hand, patients with myeloma may be hypercalcemic and suffer considerably from this symptom but still have several good years left if treated by a competent physician. In these patients, it is therefore important to know the different methods available.

For many years it has been maintained that mithramycin should be an excellent preparation for treatment of malignant hypercalcemia. We have never tried this drug since it is very toxic, but it may deserve more thorough testing in patients with highly malignant tumors where the antineoplastic effect also may be an advantage.

Calcitonin

Calcitonin in sufficient dosage, hydration with addition of active diuretics such as furosemide (Lasix), large amounts of fluids—and possibly phosphates and glucocorticoids, may be used to obtain a quick response. Calcitonin is especially effective and should be used in all serious conditions. The dosage is obviously important, whereas the type of preparation used is not so decisive. The usual preparation is salmon calcitonin, and this may be used for long-term treatment, for instance, in Paget's disease, without causing any serious antibody response. It is very potent and lowers serum calcium levels in all types of hypercalcemia due to malignancy. The effect is excellent but the mechanism is not clear.

Calcitonin also has been given for long periods to two patients who had both hyperparathyroidism and Paget's disease. Their serum calcium levels were lowered with 160 units of calcitonin per day, and the effect was much greater when inorganic phosphate was administered along with the hormone. A dose of 1500–2100 mg of phosphate was given, and it is probable that this preparation alone may have sufficient effect. However, the levels *did not return to normal before* the adenomas were removed.

Phosphates and Sulfates

The drop in the calcium level after peroral or intravenous administration of sulphates or phosphates is very rapid and the resulting changes in the patient's condition occur very quickly. This is a great advantage in a lethargic or drowsy, vomiting patient who cannot drink any fluids. On the other hand, the calcium level may drop too quickly with tetany as a result. Also the continued treatment with phosphates in parenteral administration may be dangerous from the point of view of tissue calcification. We have seen one patient in whom the treatment with intravenous phosphate was continued too long. He developed severe calcification in many organs, and I am convinced that this was of serious consequence. Therefore, it is best to administer phosphates or sulfates perorally even if it is necessary to give the first doses intravenously until the patient is able to swallow.

Goldsmith and Ingbar have successfully treated 19 patients with cancerinduced hypercalcemia with phosphates and sulfates (159). A solution containing 0.081 mole of Na_2HPO_4 + 0.019 mole of KH_2PO_4 was administered intravenously to 10 patients, eight of whom received one injection and the other two, two injections. This solution had a pH of 7.4, and one liter contained 3.1 g phosphorus and 162 mEq sodium in 19 mEq potassium. Nine patients were treated perorally with 1.5–3 grams of phosphorus per day for 7–10 days. The total doses given were rather variable, however. Only in one patient with mammary carcinoma in a desolate state did the serum phosphate level reach pathological and dangerous levels (14.8 mg/100 ml). The starting level in this patient was 2.9 mg/100 ml, but the serum calcium was 19 mg/100 ml. Even very high calcium levels were brought down to normal in most cases.

It is possible to make a more concentrated solution of phosphate and dilute it with glucose solution just before injection. One liter should be given over a 6–8 hour period. Peroral solutions may be prepared as disodium or dipotassium salts, but it must be remembered that the taste is very disagreeable, and it is therefore difficult to administer to nauseated patients. The response is rapid after both oral and intravenous administration and is clearly visible within the first 24 hours. Strangely enough the maximum effect is not reached until 1–5 days after infusion. It is thus clear that this is really an excellent way to treat cancer-induced hypercalcemia, if care is taken to avoid overdosage.

2. CALCITONIN

The question of whether calcitonin is formed ectopically has been investigated recently with the aid of the new, very sensitive methods of radioimmunoassay. In 1974 Silva et al described high levels of this hormone in the serum of a patient with oat-cell carcinoma of the bronchus (396). Coombes et al have studied this problem on a broader basis (86). They examined the sera from 28 patients with carcinoma of the breast. They found that none of the 69 control patients had raised levels. A special group of 14 patients with benign mammary lesions showed no plasma calcitonin. Out of 13 patients with "localized" cancer, only one had a high value. Metastatic cancer occurred in 28 patients. Of these, only five had normal values and a number had more than 10 times the upper limit value. The authors therefore point out "that plasma calcitonin measurements may be useful in staging patients with breast carcinoma." This practical clinical application seems most promising as a signal of cancer and metastases.

The article by Coombes et al also contains information of great theoretical interest. They were able to make monolayer cultures of the carcinoma tissue from some of their patients. In 8 out of 15 cultures from breast carcinoma cells, an immunoreactive substance was found. Three cultures continued to release hormone for 4, 6, and 10 weeks. Mammary tissue and fibroblasts in culture produced no such substance. Two out of six also released immunoreactive CEA, and three out of 12, HCG.

In another experiment breast carcinoma tissue was "cultured" in a nude mouse. The extract from the transplanted tumour contained a substance that seemed to be calcitonin.

It will be interesting to find out if the site of metastases (bone, liver) influences the degree of calcitonin increase.

In the case of osteosclerotic lesions associated with tumors, investigation of calcitonin levels may well be worthwhile. Doctors in Liège have found very high calcitonin levels in a patient with the syndrome of myeloma, osteosclerosis, and polyneuropathy. We have examined the serum from one such patient in whom the disease was probably on the wane. The serum contained increased levels.

Another relevant problem is the possible explanation of calcified—or ossified—cancer metastases. It is well known that osteosclerosis occurs in mastocytosis. We have also seen it in one patient with gastric carcinoid. It seems probable that the local effect of the tumor stimulates osteoblasts (373). The plasma levels are perhaps no true measure of production.

Perhaps it might be expected that calcitonin should cause hypocalcemia as an indication of ectopic secretion. On the other hand it is a remarkable fact that patients with very high topic excretion of calcitonin in medullary thyroid carcinoma rarely—if ever—show signs of a low plasma calcium level (see 119).

3. ACTH AND THE ECTOPIC CUSHING'S SYNDROME (HYPERADRENOCORTICISM)

In 1962 Liddle and collaborators published a paper with the title "Cause of Cushing's syndrome in patients with tumors arising from nonendocrine tissue" (290, see also 255). Three years later Liddle introduced the expression "ectopic humoral syndrome." In these studies the authors found that some patients with different malignant tumors developed Cushing's syndrome, often with hyperpigmentation. They were also able to demonstrate that ACTH-like substances could be extracted from these tumors. It was assumed that the tumors elaborated polypeptides either identical with, or similar in activity to, pituitary ACTH. A very large number of publications on this syndrome has now appeared and it is impossible to keep track of all these communications especially as they may be hidden under other titles.

It has become clear that oat-cell carcinoma of the lung occurs in about 60% of the cases. Already before the ectopic syndrome was known, it was noted that Cushing's syndrome might occur together with "carcinoma of the *thymus*." The possible connections between oat-cell carcinoma and thymoma discussed in another connection. Tumors identified as thymomas are next in frequency, followed by islet-cell carcinoma from the pancreas and carcinoids from the bronchus (286). On the whole it is remarkable that carcinoid tumors should be so common in this group, as well as in many other polyhormonal tumor syndromes. Perhaps, in some of these patients, hormone production should not be regarded as strictly ectopic biochemically even if the anatomical localization of the producing cells is abnormal.

Azzopardi and Williams have analyzed critically all the published case reports in which signs of paraneoplastic hyperadrenocorticism (14) were noted. They maintain that none of the reported cases of adenocarcinoma with paraneoplastic Cushing's syndrome is adequately documented. (It is possible that some of the adenocarcinomas had not been critically evaluated before.) They accept, as well documented, cases of thymoma, endocrine but not exocrine pancreatic tumor, carcinoid (rare from the intestine but not too rare from the bronchus) and three cases of ovarian carcinoma. They do not accept any testicular tumor, adrenal carcinoma, or pheochromocytoma but consider 11 cases of medullary thyroid carcinoma to be clear-cut.

One patient with oat-cell carcinoma had an increase in Crooke cells in the pituitary. J. H. Kennedy et al studied this problem (231). Preparations from the pituitaries from 30 controls and 68 patients with pulmonary carcinoma were all investigated by one person who did not know the diagnoses. An increased number of Crooke cells were found in 10 of 20 patients with oat-cell carcinoma. Three of seven patients with epithelial carcinomas and three of 35 of those with adenocarcinoma as well as two of the controls also had an increase in Crooke cells. However, the patients with oat-cell carcinoma definitely had the most marked increase.

Nichols et al observed a patient who had an arrhenoblastoma of the ovary with Sertoli cells (313). Clinically she had typical Cushing's syndrome with pigmentations. Her ACTH level did not drop after treatment with dexamethasone. The patient died. Postmortem examination showed large liver metastases that contained large amounts of ACTH. The adrenal glands were big and the hypophysis showed Crooke cells. The author points out that some previously published virilizing ovarian tumour patients may have had a paraneoplastic Cushing syndrome.

A very large number of other tumors also may be connected with ACTH production. Medullary thyroid carcinoma (see 494) as well as pheochromocytoma are members of the "APUD" series. A few ovarian tumors must be accepted as well as one ganglioneuroma (see also 374). On renal and prostatic carcinoma the data are not so convincing. It is important to realize that morphological investigation with a full postmortem examination to exclude the presence of other tumors as well as the clinical and biochemical data proving the diagnosis of hyperadrenocorticism are necessary. It is certainly very important to realize that tumors with such common localizations as the uterus, breast, and colon (hardly) never start production of ACTH (14). On the whole, it may be said that ACTH production by adenocarcinoma is extremely rare (372), especially by adenocarcinoma in the GI tract.

The clinical picture of Cushing's syndrome in the ectopic form may be characteristic but is usually incomplete due to a comparatively rapid development of symptoms. The patients' complaints are usually connected with the low level of serum potassium. Muscular weakness—"lassitude"—is therefore a dominating symptom, whereas kyphosis (osteopenia), striae, and moon face are not obvious. Hyperpigmentation in a patient with Cushing's disease, who has not undergone surgery previously, should always be an inducement to look for a carcinoma.

We recently observed a case in which this obvious symptom lead to careful survey of possible carcinoma sites. The carcinoma was found and treated. I have seen a patient who had the unusual combination of Cushing's disease with acanthosis nigricans as cancer signals. He was supposed to have a carcinoma with ectopic hormone production, and this interpretation was corroborated by the further development of the disease. Classical paraneoplastic external symptoms may give the direct clue to the diagnosis of carcinoma. Hypokalemic alkalosis is a common biochemical finding as well as some hyperglycemia.

The chemical connection and similarities between ACTH and MH cannot be discussed here, but it is clear that the pigmentation in Addison's disease is caused by overproduction of MSH activity together with an increase in ACTH. On the other hand, hypoadrenocorticism caused by primary deficiency of pituitary hormone production is connected with a pale skin because the patient lacks the stimulation of his pigment cells.

A number of examples illustrating the fact that the active treatment of a tumor may lead to improvement of the clinical status have now been published (125). It is said that Cushing's syndrome in a man is more common as an ectopic condition than as a classical Cushing's disease. The best proof that the tumor tissue excretes ACTH is of course the establishment of a gradient. Only with the advent of very sensitive methods of radioimmunoassay has it become possible to measure such minute amounts of hormone. Several authors have established a positive gradient, and I think that very few persons, who have studied these problems, would deny the fact that the tumor cells in these cases are the active producers of such material (198, 323, 372).

A much more difficult question to answer is the following: Does the tumour produce an ACTH molecule that is identical with normal ACTH or is the product only ACTH-like? A number of investigators have shown that the substance is immunologically identical with ACTH, but it may be that the amino acid sequence is altered in such a way that it does not change the hormone activity nor the immune reactivity. We know that the ACTH polypeptide contains aminoacids in a certain sequence. The loss of the last amino acids from this peptide does not inactivate the molecule, and it is also clear that parts of the molecule may be formed by the tumor tissue and still be active. This question is extremely interesting biologically.

Examination of the tumor and of plasma for the presence of ACTH has shown that a high-molecular fraction that would correspond to "big ACTH" is often present (153, 507). Besides this, a fraction is found that behaves in the same way as highly purified pituitary ACTH from man or pig on elution from Sephadex columns and during fractionated purification. In addition, radioimmunoassay and bioassay confirm the presence of an ACTH-active polypeptide. On the other hand it is probable that a number of similar but not identical peptides may be produced. Before the active substance has been completely purified and all contaminations have been removed, it is impossible to state the relationship between human pituitary type ACTH and other molecules with a similar activity. Some authors maintain that they have found ectopic ACTH that is of porcine and not of human type. (The same discussion is also relevant to ectopic PTH.) The fact that big ACTH molecules have been found is interesting because Yalow and Berson (507) have established that some extracts from the pituitary also contain high-molecular form ACTH (322, 323).

Hirata et al analyzed tumor specimens from four patients with an ectopic ACTH syndrome (198). Biochemically each had hyperadrenocorticism, but only one had clinical Cushing's disease. The content of ACTH in the tumors was measured both biologically and with radioimmunoassay. The cells were cultured and ACTH and beta MSH in the supernatant was isolated. Tritium labeled phenylalanine was added to one culture, and carbon 14 labeled phenylalanine to another. Column chromatography gave a peak that corresponded to that of "big" ACTH. If the big ACTH was subjected to mild tryptic digestion, active "little" ACTH was formed. This seems to be the second study so far regarding incorporation of aminoacid in ACTH synthesis.

The patient with clinical Cushing's syndrome, had had the disease for 9 years, but the clinical symptoms all disappeared when the tumor was removed. This man synthesized big ACTH.

A report of an interesting case observed for 12 years was published by Isawa et al (213).

> The patient, a 26-year-old woman, had had a primary bronchial carcinoid that was removed. Many years later she developed asthma and had increased amounts of 5-HT as judged from an HIAA excretion of 30 mg/day. She also had occasional flushing of the face but no diarrhea. The right lung was removed and contained a large tumor mass that was histologically identical with the tumor extirpated 8 years earlier. The HIAA excretion became normal. Four years later she developed a psychosis, a moon face, and trunk obesity with amenorrhea but no diarrhea. She was hypopotassemic and had an abnormal glucose tolerance test. Her corticosteroid excretion was increased. The patient died and the autopsy showed malignant carcinoid in the right lung hilus with widespread metastases. Both adrenals were hypertropic. The pituitary showed marked Crooke cell changes. ACTH was detected by radioimmunoassay both in the primary tumor and in metastatic lesions.

It seems clear that this was a carcinoid tumor that caused both the carcinoid syndrome and ectopic ACTH production. Some similar histories have been published previously. The paper contains a good reference list (see also 197).

4. RELEASING FACTORS

During the last few years a new chapter on endocrinology has been written. Extracts from certain hypothalamic centers have been investigated and active substances prepared. Their constitution has been determined and the polypeptides synthesized. The releasing factor RF for TSH is a tripeptide. The gonadotropin releaser called LH-RF is a decapeptide. Another type of activity is exerted by the somatotropin (HGH) release *inhibiting* factor (somatostatin). This is a tetradecapeptide. It has very widespread activity in different organs and possibly is also a neurotransmitter with great biological importance. Another polypeptide that is found in the nervous system and may well be a neurotransmitter is substance P. This has been investigated for a long time and its activity on the gut is well known. Nothing is known about the source of its production. Biochemically this is an undecapeptide that has now been synthesized.

The activity of TSH-RF is obviously complicated, and it is probable that it may also release prolactin and in acromegalic persons, HGH. Much work has been done on the biochemistry of the ACTH-RF. It has been very elusive, and its structure has never been identified. Vasopressin is probably a releasing factor, but its role is not yet clear. It is generally believed that the same releasing factor is active for FSH and LH, but this question is not definitely settled. Somatotropin is being actively investigated at present. Somatostatin has been found to be an inhibiting factor, and it is assumed that there is also an inhibiting factor for prolactin. Sectioning of the pituitary stalk has resulted in increased prolactin release.

One of the characteristic features of the ectopic ACTH syndrome is the fact that metyrapone administration has no inhibiting effect on the production of adrenal steroids. There have been published reports of isolated cases, however, in whom such a response was elicited. Upton and Amatruda tried to explain this fact by assuming a primary role of releasing factors in this syndrome. They were able to find two tumor patients with the ectopic ACTH syndrome who reacted in this abnormal way. From the tumors they prepared a number of peptides, smaller than ACTH. It was found that they could not be identified with vasopressin, but the quantities that could be isolated were much too small for chemical analysis (449).

This is of course a very interesting approach and search for tropin-releasing factors of other kinds should perhaps be performed in several ectopic hormone syndromes. The formation of other simple releasing factors for other pituitary tropins might explain the fact that some patients excrete complicated protein hormones that are not short polypeptides. These may be of normal pituitary origin and may be released by tumor-produced releasing factor.

5. ANTIDIURETIC HORMONES

One of the most enigmatic paraneoplastic conditions has been called inappropriate secretion of antidiuretic hormone (SIADH). In 1957 Schwartz et al observed two patients with bronchogenic tumors who suffered from the consequences of severe hyponatremia (385). Clinical symptoms were uncharacteristic, and the hyponatremia was found on routine examination. It could be demonstrated that renal and adrenal functions were normal, when determined with routine methods. The glomerular filtration rate was normal despite the fact that the urine was persistently hypertonic to the plasma. The easiest way to explain this anomalous situation seemed to be the assumption that ADH was produced in large amounts. It was known that the experimental combination of abundant fluid intake and administration of pitressin gives the same biochemical picture. The patients' condition improved on restricted fluid intake. It was pointed out that the same mechanism may be at work in other patients with pulmonary and central nervous system disease (brain damage).

Since this first observation, a large number of cases with the syndrome have been found, and it seems to be not too uncommon among patients with tumors of the lung (189). In 1972, Baumann et al (27) analyzed all the published case reports (about 120). SIADH was found to be connected most often with bronchogenic carcinoma. Solitary cases occurred with carcinoma in the esophagus, duodenum, pancreas colon, and ovary (409). Occasional patients with Hodgkin's disease also have shown the same symptoms. Patients who had no pulmonary tumor but who died later of pulmonary tuberculosis have been found to have had SIADH. This connection is quite mysterious. The electrolyte disturbances in persons with acute lobar pneumonia are of another type.

Brain damage may influence ADH secretion. The clinical picture is characterized by anorexia, nausea, vomiting, and finally great weakness. Epileptic fits and coma may occur. The critical serum sodium value seems to be 120 mEq/L and values below 100 have been observed in a few patients. It is clear that this is a life-threatening condition, if not treated properly. Many cancer patients have certainly died with these symptoms before the mechanism was recognized. Therapy is simple in principle: fluid intake should be restricted and diuresis promoted.

A number of patients with balances of sodium and water have been studied carefully, and it has been definitely proved that there is a normal excretion of aldosterone. Edema is never seen, which seems difficult to explain except by making a number of rather lofty hypotheses. One of the main arguments in favor of ADH as the chief factor is the absence of other signs of renal dysfunction. The same mechanism has been assumed to be the explanation of similar syndromes after other primary diseases. Perhaps the most important among these is traumatic or vascular disease of the CNS. There are two other small groups of patients with the same degree of hyponatremia and the same clinical symptoms. One group has acute porphyria and the other is the so-called idiopathic hyper-ADH, for which no probable cause is found. To me it seems possible that the porphyric syndrome often has to do with derangement of renal function secondary to changes in the circulation. A number of these patients have marked vomiting and azotemia with oliguria. In a recent publication, Eales and his group in Cape Town have interpreted the findings in acute porphyria as primarily caused by vascular processes in the kidney. It is probable that there may be many mechanisms at work producing the same effect (295).

Cutting observed a patient who developed the syndrome after intravenous massive therapy with vincristine and who obviously suffered from a severe polyneuropathy along with the hyponatremia. Some authors believe that the polyneuropathy and the SIADH are connected in acute porphyria. However, we have seen some patients with severe hyponatremia during a porphyric attack who have not developed any marked symptoms from the peripheral nerves. In our experience these two conditions are separate. One is caused by water overload and disturbed renal circulation (plus ADH?), the other (pareses) by barbiturates and other drugs causing induction of ALA synthetase. I have mentioned these other conditions to show how complicated these mechanisms really are. The crucial experiment would be to demonstrate vasopressin in urine, and in tumor tissue or, still more convincing, as a difference—a gradient—in the arterial and the venous blood from the tumor. There have been a number of positive reports stating that a substance identical with or very similar to arginine vasopressin is present in the urine and also in tumor tissue. Perhaps the most convincing experiences were published by Walter as well as by Baumann et al, and it seems hard to deny that some substance close to ADH is secreted in such patients (27, 481). Some patients were found to produce arginine-vasopressin, or a substance that reacted identically on radioimmunoassay and on bioassay. Tissue from a normal lung, lymph nodes, and lung tumor from other patients did not show any detectable activity. Extracts from normal pituitaries contained about 10 times as much hormone as the lung tumor. It will be necessary to analyze the active material more closely and make chemical determinations of amino acid sequences to confirm that this substance is identical with normal ADH.

Several other patients have been observed with lasting normalization of their water metabolism after resection of the tumor (442). Recurrence of the syndrome together with return of the tumor has also been observed. Thus it cannot be denied that some substance from the tumor is the cause of this derangement in electrolyte handling.

It is clear that a few patients have had an increased production of substances that are similar to, or identical with arginine-vasopressin. This fits in well with the concept of derepression. On the other hand, there are data that seem to be difficult to explain only by overproduction of ADH (see 177). Regarding the presence of neurophysin see 169.

An interesting paper was published from St. Thomas Hospital in London (22). Six patients with anaplastic oat-cell carcinoma of the bronchus were observed. They had symptoms associated with overhydration and hyponatremia. The excess of ADH-material was assayed (Lee's method) and the results suggest that it is vasopressin although the tests cannot specifically identify it as the human arginine-vasopressin. Urines from six patients and sera from four were investigated. In five of six cases the substance was found in the primary and/or secondary tumor. No mention is made of the total number of persons with bronchial carcinoma that were investigated. Plasma sodium values ranged from 110 to 129 mEq/L. Despite subnormal plasmatonicity, the urines were hypertonic. None of the patients who received a water load was able to excrete it normally. Total body water was increased in all four cases with such analysis. No patient had evidence of renal impairment. Five cases had extensive tumor spread. The antidiuretic principle may be a related but not identical polypeptide (22).

Tisher observed a 55-year-old woman with bronchogenic carcinoma. She had persistent urine hyperosmolality with hyponatremia and a normal blood volume. The hyponatremia was corrected by fluid restriction. The patient had normal renal and adrenal function. All symptoms dispppeared after tumor resection. The tumor and hilar metastases showed antidiuretic activity corresponding to 120 and 40 USP units of arginine-vasopressin, respectively. Recurrence of SIADH preceded relapse of the tumor. At autopsy tumor tissue was tested and found to contain high amounts of ADH (442).

Baumann et al published the results of an excellent study of the problem made at Peter Bent Brigham Hospital in Boston (27). The authors devised a way to concentrate the antidiuretic factor and used a standardized bioassay method to determine the amount of the antidiuretic factor present. Inactivation with thioglycolate was also tested. They found that most arguments for a real paraneoplastic increase of ADH are rather indirect. Four patients with oat-cell carcinoma, two with tumor cerebri, and one with acute intermittent porphyria were supposed to have SIADH.

The treatment of the acute situation in SIADH is quite rational—it consists of fluid restriction and use of potent diuretics. The best results seem to have been obtained with furosemide. Radical extirpation of the tumor is of course the method of choice if practicable.

6. COMPLETE GONADOTROPIN MOLECULES

French clinicians pointed out long ago that gynecomastia may occur in patients with lung tumors. The first publications regarding this question are difficult to locate, but in 1915, in a paper on osteoarthropathy, Locke mentioned the fact that enlargement of the breasts had been seen in a man with a pulmonary neoplasm. This paper treats osteoarthropathy in an extensive and excellent presentation. Boudin (57) writes about a syndrome in which gynecomastia is combined with osteoarthropathy, and he notes that Bariéty and Cury (20) studied 25 cases (in 1950), some with this combination and others with only one of the symptoms.

Fusco and Rosen were the first to realize that gynecomastia could be explained by the fact that the tumor produced gonadotropins (147). They found four patients with this symptom and described the tumor histology as anaplastic, large-cell carcinoma. It is clear, however, that several later patients have had adenocarcinoma or oat-cell carcinoma and many have had squamous cell carcinoma. In many cases gynecomastia has been the most obvious symptom, and there have been several observations of its disappearance after extirpation of the tumor. An unusually striking case report was published by Faiman et al (134).

> The patient, a 49-year-old man, had gynecomastia, severe pains in knees and joints, exudate in the right knee, clubbing, and a typical periosteal thickening. His right lung was removed. The tumor was mainly an adenocarcinoma but pleomorphic and epidermal parts were also noted. The joint symptoms disappeared rapidly, and the gynecomastia was no longer seen after 3 weeks. There was an arteriovenous difference of radioimmunologically determined FSH. The levels in blood and urine became normal when the clinical symptoms disappeared.

Another patient was described by Dailey and Marcuse (99).

> This man had gynecomastia and signs of a lung tumor. His urine contained very high levels of chorionic gonadotropin (320 IU/ml) as did the tumor, which was a giant-cell anaplastic carcinoma. The testes showed marked hyperplasia of the intestitial cells but no signs of chorionepithelioma could be found. The authors believe that the hormone was chiefly of a nonpituitary, i.e., of a chorionic, type with chiefly luteotropic properties.

Cottrell and Becker observed four patients with bronchogenic carcinoma and increased gonadotropin levels (93). All tumors, although histologically different, contained considerable amounts of hormone. The tumor in the first case produced keratin. In the second case there was no keratinization but the metastases showed signs of squamous cell carcinoma and adenocarcinoma. The tumor in the third case had small round to oval cells. In the fourth case, the tumor was difficult to classify but some cells contained mucin (see also 32).

A recent paper shows that also the kidney may be the site of gonadotropin producing tumors (158). The first patient described was a man, aged 53, who had an abdominal mass, left-sided gynecomastia, small testes, and an increase in serum alkaline phosphatase and urinary gonadotropins. A clear-cell renal carcinoma was extirpated and the gonadotropin levels decreased.

There is one special tumor that produces ectopic gonadotropic hormones, and this is liver cell carcinoma in youngsters. All patients with precocious puberty caused by this tumor have been boys. This is probably due to the fact the testis needs only FSH to develop, whereas the ovary needs both FSH and ICHH. Girls with hepatoma may have high levels of ICHH but never combined with FSH. As far as could be ascertained, there were no admixtures of choriocarcinoma. No case of feminization has been described in this group (209, 271, 362).

Braunstein et al have published a report of an interesting observation of hepatoblastoma in a 4-month-old boy (61). There were signs of pubertas praecox. HCG and fetuin levels were high but diminished after tumor resection when the boy was 11-months-old. The levels rose again when the patient developed lung metastases and then showed increased virilism. The testes showed no trophoblastic tissue. This is an interesting instance of a tumor with the production of *two* proteins. The fact that the fetuin was produced is probably explained by the patient's age (The hepatoma probably was present at birth.) Tumor tissue was cultivated in vitro and *produced both HCG and fetuin* over a 3.5-month period. (See also 62).

Our knowledge of the nature of HCG is incomplete since it has been difficult to distinguish with certainty between LH and FSH. It may be important to remember that there are patients with increased gonadotropin levels and gynecomastia without increased estrogen levels. It will certainly be of interest to find out if mammotropic hormones are also excreted in some patients. K. L. Becker et al saw a man of 74 with bilateral gynecomastia but no milk secretion (32). There was no clubbing. A large nonresectable lung tumor was present (bronchogenic epidermoid). Gonadotropin values were equivalent to those found in the early weeks of human pregnancy, and the concentrations in the lung tumor and in metastases were very high. The hormone was demonstrated with the aid of immunofluorescence. Treatment with testosterone, cortisone, and stilbestrol did not influence the titer. The testes were examined carefully and were found to contain no choriocarcinoma. In this patient no estrogen fraction was found to be increased. Levels of plasma insulin, growth hormone, and thyrotropin as well as serum calcium were normal. It is remarkable that gynecomastia should occur without elevation of the estrogen level.

Turkington observed a woman who had galactorrhea and a hypernephroma with high serum prolactin levels. Excision of the tumor produced a rapid drop in the hormone level. The tumor cells were cultured and released a substance with a biological prolactin effect. This was neutralized by the addition of specific antiserum that did not react with HGH. The same author observed a man with a lung tumor, and no milk secretion. After the tumor was irradiated, his high prolactin levels decreased markedly (446).

Fusco and Rosen demonstrated ectopic human placental lactogen (HPL) in 10 of 128 sera (8.P) from patients with nontrophoblastic cancers. All these patients had gynecomastia and elevated estrogen levels (147). It seems probable that these patients also had elevated HCG levels (see also 328).

Oat-cell carcinoma is not always connected with one special hormone product. It is evident, however, that oat-cell carcinoma in the lung has a very wide spectrum of hormone synthetic capacity. This is true both for ACTH and ADH production. PTH seems to be largely connected with squamous cell carcinoma and FSH with mixed types. On the whole it is probable that multiple production of hormones is quite common in lung tumors and we shall see many more combinations, when we are able to define spectra of hormone patterns, e.g., with radioimmunologic methods (see 317).

Pineal tumors may produce precocious puberty but only in boys. Forty-six out of 145 boys with such tumors showed early sexual development. The probable explanation is that melatonin—a normal gonadotropin inhibitor—is not produced because the normal pineal gland is destroyed. Therefore, this is a mechanical not a humoral, truly paraneoplastic syndrome (506). Other explanations are possible.

7. GLYCOPROTEIN HORMONES AND THEIR SUBUNITS

Both the gonadotropin and the thyrotropin molecules contain two polypeptide chains. One is a beta subunit and the other one, alpha. We have reason to believe that the alpha subunits in the glycoprotein tropic hormones TSH, LH, FSH, and HCG are identical, whereas the specific action is connected with the beta unit. Very recent experiments have shown that stimulation of luteinizing hormone (LH) release with the specific releasing factor (HRF) produces a rapid increase in alpha chains after 20 minutes with a somewhat later increase in the peak of complete LH. The relationship between excreted LH and alpha chains is 4:1, if the dose of LH-RF is sufficient. Experiments prove that the alpha chains do not appear as breakdown products of LH, and that injection of intact LH does not give rise to any free alpha chains. This seems to indicate that alpha chains may be synthesized and excreted separately just as light chains (Bence Jones protein) may be excreted in some patients with myeloma.

Hagen and McNeilly have shown that TSH-RF may also cause an isolated initial release in alpha chains after 20 minutes with a marked increase in the whole TSH hormone following (175). TSH-RF gave no increase in the beta chains of LH. It is thus clear that the alpha chain may be secreted isolated, and also that the specificity of the releasing hormone is only limited to the excretion of the whole polypeptide hormone molecule.

Rosen and Weintraub have been able to show that there might be imbalanced production of HCG alpha units as well as isolated production of the beta subunit by nontrophoblastic cancers (365). In a paper from 1974 these authors reported the results of an investigation of the excretion of alpha subunits in a patient with cancer whose case history was published previously by Engelman et al (126).

A 27-year-old woman had had an ulcerated gastric carcinoid tumor that was resected. She had a blotchy flushing (typical of a gastric carcinoid syndrome) and also vomiting and diarrhea. An ulcerated gastric carcinoid tumor was resected but she had numerous peritoneal and hepatic metastases. Her diarrhea reacted very well to parachlorophenylalanine but recurred when she was given a placebo; it disappeared again when therapy was reinstituted.

HIAA production was not very high but, typically, she had a high 5-HTP level. In 1967 it was noted that radioimmunoassay showed indications of large amounts of gonadotropins. At that time nothing was known about isolated alpha subunits. When it later (1974) became possible to use specific antiserum against alpha chains, her serum was found to contain amounts more than 1000 times higher than in any sera from patients without cancer. The amount of total gonadotropins was much lower and, as a matter of fact, was not pathological. (This patient is probably the most multihormonal that has been described. She obviously produced a large number of active substances. This pattern is discussed in this chapter). In this connection it was found that her serum FSH level was increased almost twofold, whereas her LH and TSH levels were normal. She had considerable hypercalcemia, (11.7 mg of calcium per 100 ml), with a slightly increased PTH level and a much increased calcitonin level. Gastrin concentration was low. The tumor contained a considerable amount of PTH, and some increase in calcitonin, and had a marked elevation of alpha units.

The alpha chain purified from the tumor showed definite differences between tumor alpha and the standard alpha subunit. During the period when she was treated with parachlorophenylalanine, 5-HT production was much inhibited, whereas this medication had no apparent effect on the serum alpha unit concentration.The patient was also treated with Melphalan, and this appeared to halt the rate of rise in serum alpha units. When the drug was stopped, this was followed by a rapid rise in serum alpha units. The authors have also seen a case of isolated production of the beta subunit of HCG (365).

The same authors have investigated serum alpha units in 131 patients with cancer but without indication of trophoblastic disease. Of 106 sera from patients with bronchogenic carcinomas, three were positive. They also examined sera from a large number of patients with increased gonadotropin or TSH levels from different causes. None had increased alpha chains (365, see also 118). Ectopic production of somatomammotropin has also been established in nontrophoblastic cancers (487).

Braunstein et al have performed a very extensive investigation of HCG in the blood of cancer patients (63). They studied sera from 828 patients with nontrophoblastic or nontesticular tumors with a very sensitive radioimmunologic method for the specific quantitation of the beta subunit of HCG. The pituitary LH levels were not determined. A surprisingly high number (60) of persons with varying location of malignant tumors was found to have increased amount of HCG. The hepatomas were—as expected—the leading tumors (14/81) but gastric carcinoma (8/34), pancreatic carcinoma (5/10), as well as melanoma (8/77) were high on the list. Sera from some patients with lymphoma also gave positive results—2/69 from patients with lymphosarcoma and 3/147 from those with Hodgkin's disease belonged to the positive group. Four sera of 65 from patients with myeloma also gave positive results. Unfortunately, no individual data were given. The last named disease occurs in the higher age groups and the presence of a second malignant disease is not uncommon. This has to be excluded but the fact that plasma cells might be ectopic producers of HCG is remarkable and should be investigated more closely.

The authors were obviously impressed by the verdicts of the pathologists. They cited publications indicating that there may be a "morphological retrodifferentiation" of an adenocarcinoma to a choriocarcinoma. Derepression of polypeptide synthesis is of course a functional "retrogression" that seems to be more reliable than speculations built on morphologic observations of "resemblances."

8. OSTEOARTHROPATHY

One of the best known external changes connected with a neoplastic process is the so-called hypertrophic osteoarthropathy–h.o.a.—(ostéoarthropathie hypertrophiante pneumique). This means not only clubbing of the fingers, which is in itself a common finding in many chronic or subchronic pulmonary or cardiac diseases. Real osteoarthropathy should be characterized not only by convex nails ("watch glass"), and hypertrophy of the end phalanx, but also by swelling of the hands, and periosteal calcified thickening of the lower parts of ulna and radius (see 218). Diagnosis therefore should not only be founded on simple observation alone but must be strengthened by radiology. The findings on the lower legs are less constant and sometimes more difficult to evaluate. It should be remembered that periosteal calcification also may occur in the lower part of the femur.

The patients' complaints are often very severe. They may give a history suggestive of subacute rheumatoid arthritis. Such a history is of course important from the practical point of view as arthritis has been an incorrect initial diagnosis in several cases. After radical surgery all the subjective complaints disappear as well as the marked swelling of the hands. It may be that this part of the syndrome should be studied more closely since it may be an important parallel to chronic rheumatoid arthritis.

The literature contains reports of observations indicating that the same symptoms may also be seen in other intrathoracic maladies such as chronic infections, but the most classical picture is connected with lung tumors where it may be of real diagnostic importance even though the phenomenon is uncommon. The combination of osteoarthropathy and gynecomastia—sometimes unilateral—makes the suspicion of an intrathoracic neoplasm very strong. It seems to be a common finding that pain indicates the presence of tumor, whereas other pulmonary diseases only cause swelling.

An excellent review of h.o.a. has been published by Hammarsten and O'Leary (180). These authors collected no less than 22 cases of their own. All of their 22 patients showed changes in the hands and wrists on X-ray films. Twenty patients had malignancy of the lung. One had an infected infarction of the lung, the other had pulmonary fibrosis. The most common first diagnosis was rheumatoid arthritis and the pulmonary neoplasm was often overlooked, despite the fact that many of the patients were seen by numerous physicians. Ten of the 21 men had gynecomastia; eight had acromegaloid features. The two patients, who had no malignancy, had no joint pain. This was a striking symptom in the tumor patients. In seventeen patients joint symptoms were the first or only evidence of disease. In many patients both legs and forearms showed diffuse thickening without soft tissue edema. All patients, who had the primary tumor resected, showed dramatic relief in 24 hours or less. The changes seen radiologically disappeared very slowly but never completely. Six patients were treated with cortisone which substantially relieved the joint pain, and this obviously confused the diagnosis. There was no correlation between the presence of gynecomastia and abnormal liver function.

Ray and Fischer studied cases with h.o.a. in both pulmonary suppuration and in carcinoma. There were no arthritic symptoms in the first group, whereas these were severe in the tumor patients.

Presently, the mechanism is unclear. Certain parts of the syndrome resemble acromegaly. I have personally seen patients with the latter disease who have had extremely swollen, severely painful hands. After surgical extirpation of the pituitary adenoma, the swelling of the hands disappeared and the patients were able to put on their rings as usual. These changes may occur in one to two weeks, and are very similar to those seen after thoracic surgery in patients with a lung tumor. On the other hand thickening of the periosteum is not a characteristic of the acromegalic process.

Remembering the parallels with acromegaly, we investigated the levels of human growth hormone (HGH) in one patient who had this disease, and we found increased values (412). It is possible that this finding can be interpreted in several ways. We have observed two additional cases. In one of them the clinical symptoms disappeared and the levels of the growth hormone became normal after operation. The tumor in the other patient was inoperable.

The literature does not contain many reports about further investigation of this interesting condition. Greenberg et al observed the formation of labeled HGH by cultured cells from such a patient with carcinoma of the lungs (169). Other authors have had more variable results. Sparagana et al studied the presence of HGH in the tumor tissue and in the surrounding lung (410). The only case with h.o.a. had the highest values. Another study was performed by Dupont et al from Denmark. They found that one patient with h.o.a. had high fasting HGH levels, whereas two other patients did not show any such signs. In a recent study Beck and Burger working in the same hospital as Cameron et al extracted HGH from bronchial and gastric carcinomas and found that seven of 18 in the former group and five of eight in the latter had increased contents (30). Anaplastic tumors had the highest concentration. Three of the seven patients, whose lung tumors contained a relatively high concentration of HGH, had typical osteoarthropathy, but this was also found in one of the patients with low levels.

It seems clear to me that the production of HGH by anaplastic carcinoma cells has been definitely proved, but the development of osteoarthropathy is in many cases independent of HGH when this is determined by radioimmunoassay. These problems thus await further studies. We should not forget that vagotomy may have strong effects on the acute symptoms of osteoarthropathy in patients with lung tumors. Evidently a complicated pattern of different factors must be responsible (206).

Tumors with other localizations may also cause h.o.a. (327, 445).

> Hollis studied a patient who had typical osteoarthropathy and "rheumatic" pains in her elbows, shoulders, and knees. Her fingers were very swollen and painful on movement. There was typical clubbing, and it was remarked that her coarsened facial features were suggestive of acromegaly. A polypoid tumor in the gastric fundus was found and extirpated. It was a myxoma weighing 78 g and attached by a long pedicle to the stomach wall. Macroscopic sections showed a cellular tumor composed of uniform stellate cells embedded in a myxoid stroma; no anaplasia was noted. Within 24 hours after surgery the pain subsided, and after 48 hours her hands and feet were smaller. Her skin appeared shriveled. On the seventeenth day all postoperative swelling and pain were gone. She developed a pulmonary embolism and died. No other tumors were found at the postmortem examination (202).

The author states that 5% of all patients with carcinoma of the lung have such symptoms. Fifty percent of all patients with diaphragmatic neurilemmoma are said to have h.o.a. Tumors in the upper GI tract are very rarely present. There have been published reports of osteoarthropathy in one patient with carcinoma of the esophagus, in one with carcinoma of the stomach, and in three with carcinoma of the nasopharynx. In a few cases subsequent radiological examination has shown that the condition of the bone may return to normal in 6–9 months after operation. Histological studies of the bones show successive layers of bone formation. Even if the combination of gynecomastia and changes in the hands are common, there are patients with lung tumors who have only one of these symptoms.

French authors have been very interested in this syndrome, and they talk about "dysacromelie," a term proposed by Barièty and Coury who described 25 cases in a series of papers in 1950. The syndrome may consist of pure hypertrophy of the distal parts of the fingers, or isolated periosteal changes, or a real polyarticular syndrome that mimics rheumatoid arthritis. These authors also discussed the syndrome in combination with gynecomastia, but it is difficult to find any data on the frequency of this complication and its combination with osteoarthropathy.

9. VASOACTIVE INTESTINAL POLYPEPTIDE (VIP) AND "PANCREATIC CHOLERA"

A special type of case was separated from those with the general Zollinger-Ellison syndrome when it was pointed out that diarrhea may occur in the absence of gastric hypersecretion. Previously it had been assumed that the hypergastrinemia induced such a profuse secretion of fluids that there was an "overflow" through the colon. The syndrome has been called pancreatic cholera and a number of such cases are known. The syndrome consists of copious watery diarrhea, hypokalemia (obviously explained by the rapid transit time through the colon), and according to some authors, also achlorhydria (173). The responsible chemical agent has been sought in several classes of molecules: at first gastrin, then glucagon, secretin, prostaglandins, and so forth. Bloom et al investigated six patients and reported elevated plasma levels or high tumor content of the so-called vasoactive intestinal polypeptide (VIP) (50). (Gastric inhibitory peptide [GIP.] also has been regarded as a cause of severe diarrhea). Further case reports have been published in which Said has found an increase in VIP in plasma, tumor, or both. The number of such cases observed by Bloom and by Said is quite large, but it is difficult to find a close connection between VIP values and clinical symptoms.

One of the most remarkable cases of biochemically induced diarrhea connected with the presence of a tumor is the following reported by Murray (308).

> The patient was a 49-year-old man. In 1943 he noticed the sudden appearance of pronounced flushing in his face and upper trunk together with a throbbing generalized headache lasting 15 to 30 minutes and followed by copious watery brown stool. This occurred 5 to 10 times a day. The diarrhea was refractory against treatment. On admission the pronounced flushing was observed but not described in detail. There was no gastric hypersecretion during an attack of diarrhea. The stools were enormous with one single evacuation of 1300 ml and a maximum of 7.8 liters in 24 hours! Biochemically hypokalemia, metabolic acidosis, and mild azotemia were present. HIAA level was normal. The catecholamine levels were at the upper limit. After administration of large amounts of potassium, a laparotomy was performed and a tumor, weighing 132 grams, was extirpated from the tail of the pancreas. The microscopical picture varied considerably in different parts of the tumor. It was

> partly glandular, partly embryonal. Gomori staining showed no granules. A needle biopsy specimen from the kidney showed tubular changes. After the operation the attacks disappeared. The patient was free from symptoms during a 14-month observation period.

This was obviously an atypical Zollinger-Ellison syndrome.

10. MULTIPLE HORMONE PRODUCTION

Some tumors have been found to produce several polypeptide hormones. It is of course difficult to decide if this production is really ectopic since many examples are carcinoids or oat-cell pulmonary carcinomas. These are obviously very multipotent and it is possible that cells constituting these tumors are derived from a stem cell belonging to the neural crest system.

In 1965 Liddle found two patients with pulmonary oat-cell carcinoma producing ACTH and MSH. When we remember the close connection between these two molecules, it is not easy to decide if they should be regarded as different hormones. One of the patients, however, also had hypercalcemia and was regarded as having an increased production of PTH. Later a large number of oat-cell carcinomas were found to produce both ACTH and beta-MSH, and this is obviously a common combination (248).

> Rees et al (349) described a patient who complained of swelling of the face and ankles, hirsutism, pigmentation, and weakness. She had a cushingoid habitus and a pulmonary tumor. It was impossible to remove the tumor radically, but parts were examined and diagnosed as oat-cell carcinoma. This operation produced some regression of her symptoms. Later, her liver became enlarged, she developed hyponatremia, and died. At autopsy widespread metastases of a small cell pulmonary tumor were found. ACTH and beta-MSH were much increased. Radioimmunoassay and bioassay showed that Arginine-vasopressin was also increased in the tumor. Neurophysin, oxytocin, and prolactin were much increased when determined by radioimmunoassay.

The authors point out that oat-cell carcinoma may be of endocrine origin, possibly arising from what the German pathologist Feyerter called "Helle Zellen" (light cells) in the bronchial mucosa. It is important that secretory granules seen on electron microscopy are found in carcinoids *and* in oat-cell carcinomas. The tumor was also investigated for the presence of glucagon, LH, gastrin, and HGH but no pathological substance was found in the tumor extracts. This observation might indicate that these tumors also produce multiple hormones that do not give any clinical symptoms.

In a number of patients with oat-cell carcinoma neurophysin has also been present. The combination of AVP and oxytocin and/or ACTH has been found in several oat-cell carcinomas and in one malignant bronchial carcinoid. Marks found a patient with adenocarcinoma of the pancreas that contained both AVP and oxytocin. In Malmö we have seen a patient who had the typical carcinoid picture with flushing and also severe hypoglycemia. Histologically the tumor was an adenocarcinoma of the pancreas.

Faiman et al and Weintraub and Rosen studied adenocarcinomas and undifferentiated carcinoma of the lung for the presence of gonadotropins. In one case they found FSH and LH, in others, HPL and HCG.

On careful search in the literature, a number of other such observations are found. A paper by O'Neal et al (320) contains data from one patient with a mediastinal tumor that produced ADH and ACTH. One carcinoma of the lung and one of the pancreas produced both ADH and PTH. A patient with pheochromocytoma producing topic noradrenalin and ectopic ACTH is also of interest. The islet cell tumors may contain both insulin and glucagon as well as a number of other active substances.

In 1965 Choh Hao Li published the results of his investigations on a pituitary protein that he called beta-lipotropin (beta-LPH). This is a rather big molecule containing 90–91 amino acids. Different parts of the molecule are identical with certain factors that have a high activity. The so-called melanocyte stimulating hormone (beta-MSH) is contained within the 37–58 sequence. It is therefore probable that beta-MSH is a split product from beta-LPH. On the other hand, alpha MSH is identical with the 13 amino acids at the terminal end of the ACTH molecule. Therefore it is important to decide if the so-called ectopic ACTH syndrome and its heavy pigmentation is really caused by production of larger prohormones. During the past few years much interest has been generated by the discovery that the beta-LPH molecule contains sequences that are identical with beta-endorphin. This pentapeptide also may be obtained from beta-LPH, and it is probable that it may be a strong releaser of HGH, in man as well as in the rat. In the rat it also releases prolactin. Its analgesic morphinelike activity is presently being actively studied in many centers, e.g., at the NIH by Nirenberg. Many observations seem to favor the assumption that such polypeptide fragments also could be produced ectopically by cancer cells, and thus they may become of great importance for oncology (115).

At the 1977 meeting of the Association of American Physicians, Odell et al presented their data on ectopic peptide synthesis as a universal concomitant of neoplasia. These authors had performed radioimmunoassays or receptorassays of ACTH, big-ACTH, beta-lipotropin (beta-LPH), human chorionic gonadotropin (HCG), alpha units of HCG and vasopressin-vasotocin. Extracts from a number of different carcinomas, chiefly of the lung and colon, from patients with no ectopic hormonal syndromes were analyzed and were found to contain one or several of these substances. In 60 other patients with untreated lung cancer, peripheral blood samples showed elevated levels in different percentages. The same was also true for sera from patients with cancer of the colon. Peripheral blood HCG and beta-HCG levels were elevated only in 5% of these patients. In many blood samples, the elevations were quite marked, and if the battery of tests is combined, a very high percentage of the sera shows one or several pathological values. It is therefore possible that such tests may become valuable diagnostic tools in the future. This was discussed in another paper on the use of peptide markers for early diagnosis of lung cancer. The data given were most interesting, and some observations indicated that positive findings in three of 10 patients, who were regarded as suffering from only chronic obstructive pulmonary disease may have indicated early carcinoma.

Ducray has written an excellent review of the molecular evolution of gut hormones. He points out that the structures of the different "hormonal families" have been developed during the last decade. There are important homologies between the polypeptide in gastrin and in cholecystokinin (CCK) and also another related polypeptide coerulein. It has been known for some time that secretin and glucagon share a sequence of 14 amino acids, and it is possible that there is a homology between other intestinal polypeptides (vasoactive intestinal peptide [VIP.] and gastric inhibitory peptide [GIP.]). There are also a number of other peptides with biological activity produced by the intestinal cells or possibly by the liver. Substance P is one of these. It has been found in some patients with carcinoid tumors (see the Appendix.)

The fact that several of these peptides occur in different sizes is also a proof of the great variability. Big gastrin has 34 amino acids and gastrin, 17. Gastrin-producing tumors seem also to contain 14 amino acids. A miniform of CCK is also known in two forms, one with 33 and another with 39 amino acid residues. Inactive forms also have been isolated, and this seems to indicate that we may expect a large number of different individual peptides that may be synthesized both topically and probably also ectopically.

Recent investigations have shown that the large β-lipotropin molecule contains a number of active amino acid sequences. Proteolytic activity probably liberates these different active fractions, and it seems well worth thorough investigations to find such fragments in patients with carcinoma. This very polymorphic situation may account for the fact that immunological methods and measurement of biological activity sometimes give different results. The immunoreactivity of COOH and NH_2 terminal parts of the molecule may be quite different.

The number of patients in whom increased amounts of different peptides probably produced by carcinomas have been found has increased rapidly with improving techniques. One of the most important publications is a recent one by Coombes et al (86). These authors found increased levels of calcitonin in a large number of patients with different tumors. Braunstein et al (61) have investigated the presence of free alpha chains from pituitary glycopeptide hormones in a large group of patients with different carcinomas and have had many positive results. In all probability cancer cells may produce a variety of different polypeptides as part of the disturbed protein synthesis. The presence of such products can only be suspected when active molecules occur in such a high concentration that they have clinically visible results.

Appendix 1. The Most Convincing Results of Investigations on Hormone Production by Cancer Cells in Vitro

Braunstein et al (61)	HCG + fetuin
Coombes et al (86)	Calcitonin
	Calcitonin + CEA
	Calcitonin + HCG
George et al (152)	ADH
Greenburg et al (169)	PTH
Hirata et al (198)	ACTH
Klein et al (237)	ADH
Orth (322)	ACTH
Rabson et al (346)	HCG
Turkington (446)	Prolactin

Observations on a gradient over the tumor:

Balsam et al (18)	ACTH	Adenocarcinoma coli
Blair et al (45)	PTH	Renal
Buckle et al (70)	PTH	Renal
Faiman et al (134)	HCG	Adenocarcinoma pulm
Knill-Jones et al (239)	PTH	Liver
Ratcliffe	ACTH	Lung
Schteingart et al (383)	ACTH + β-MSH	Paraganglioma

Appendix 2. Molecular Characteristics and Source of Hormones and Other Specific Substances

Hormone or Substance	*Mol. W. (Daltons)*	*Amino Acids (Number)*	*Source*
Aldosterone } Steroids	288	—	Never truly ectopic
Testosterone } Steroids			
Estrogen } Steroids			
TSH-RF } Releasing factors		3	Possibly ectopic
ACTH-RF } Releasing factors			
LH-RF } Releasing factors		10	
Endorphin (from beta-lipotropin)		5	?
Oxytocin	1007	9	TRM 3
Vasopressin (ADH)		9	Ectopic
Substance P		11	?
Somatostatin		14	Pancreas—D-cells, etc
Gastrin	2200	17 (4-5 active)	Zollinger
Minigastrin		13 (c-terminal)	Ellison
Big gastrin	4400	34	Pancreas
Beta-MSH		22	Ectopic
Secretin shares one sequence of 14 aa [a] with glucagon)	3055	27	Pancreas
Glucagon		29	Pancreas
VIP		28	Pancreas
Calcitonin	3600	33	Lung, etc.
Pancreatic polypeptide		36	Ectopic
Cholecystokinin (CCK) (five c-terminal aa [a]=gastrin)	3919	39	?
ACTH	4550	39	Ectopic
Alpha MSH (13 first aa [a] in ACTH)			Ectopic
GIP	5105	43	
Somatomedin	∓5000		
Enteroglucagon	7000		Kidney (ectopic)
PTH	9100	84	Many tumors (ectopic)
Pro-PTH	∓12,000	109	
Prolactin	13,000	92	Ectopic?
Beta-lipotropin		90	
HGH	∓21,500	191	Lung etc.
FSH, alpha unit	13,600	92 } 207	Ectopic
FSH, beta unit	∓28,000	115 } 207	
TSH, beta unit	14,700 } 28310	113 } 209	
TSH alpha unit	13,600 } 28310	96 } 209	

Appendix 2. Molecular Characteristics and Source of Hormones and Other Specific Substances *(continued)*

Hormone or Substance	*Mol. W. (Daltons)*	*Amino Acids (Number)*	*Source*
Neurophysin	∓30,000		
alpha-HCG	13,600	92 } 231	Ectopic
beta-HCG		139 }	
Placental lactogen	19,000		Ectopic?
Erythropoietin	45,800?		Ectopic
Albumin	65,000	ca 600	—

[a] aa = amino acids.

References

1. Agrup, G., Falck, B., Kennedy, B.-M., et al: Dopa and 5-S-cysteinyldopa in the urine in healthy humans. *Acta Dermatovener (Stockholm)* 53:453–454, 1973. (See also 5-S-cysteinyldopa in the urine of melanoma patients, *Acta Dermatovener (Stockholm)* 57:113–116, 1977.
2. Aguayo, A., Thompson, D. W., and Humphrey, J. G.: Multiple myeloma with polyneuropathy and osteosclerotic lesions. *J. Neurol. Neurosurg. Psychiat.* 27:562–566, 1964.
3. Albright, F. and Reifenstein, E. C.: *Parathyroid Glands and Metabolic Bone Disease.* Williams & Wilkins, Baltimore, 1948, p. 93.
4. Alexander, P.: Foetal "antigens" in cancer. *Nature* 235:137–140, 1972.
5. Amatruda, T. T., Jr.: Nonendocrine secreting tumors. *Duncan's Diseases of Metabolism,* ed. 6. Saunders, Philadelphia, pp. 1227–1244.
6. Ammann, R. W., Berk, J. E., Fridhandler, et al: Hyperamylasemia with carcinoma of the lung. *Ann. Intern. Med.* 78:521–525, 1973.
7. Anderson, E. E. and Glenn, J. F.: Cushing's syndrome associated with anaplastic carcinoma of the thyroid gland. *J. Urol.* 95:1–4, 1966.
8. Andrasch, R. H., Bardana, E. M., Jr., Porter, J. M., et al: Digital ischemia and gangrene preceding renal neoplasm. *Arch. Intern. Med.* 136:486–488, 1976.
9. Aronsen, K. F., Torp, A., and Waldenström, J. G.: A case of carcinoid syndrome followed for eight years after palliative liver resection. *Acta Med. Scand.* 199:327–329, 1976.
10. Åström, K.-E., Mancall, E. L., and Richardson, E. P., Jr.: Progressive multifocal leukoencephalopathy. *Brain* 81:93–111, 1958.
11. Axelsson, U., Hägerstrand, I., and Zettervall, O.: Unusual pattern of hepatic alkaline phosphatase activity and renal carcinoma. *Acta Med. Scand.* 195:223–225, 1974.
12. Azar, H. A.: Amyloidosis and plasma cell disorders, in Azar, H. A. and Potter, M. (eds.): *Multiple Myeloma and Related Disorders,* vol. 1. Harper & Row, New York, 1973.
13. Azzopardi, J. G. and Whittaker, R. S.: Bronchial carcinoma and hypercalcemia. *J. Clin. Path.* 22:718–724, 1969.
14. Azzopardi, J. G. and Williams, E. D.: Pathology of "nonendocrine" tumors associated with Cushing's syndrome. *Cancer* 22:274–286, 1968.
15. Bagshawe, K. D.: Tumour-associated antigens. *Brit. Med. Bull.* 30:68–73, 1974.
16. Baker, L. R. I., Brain, M. C., Azzopardi, J. G., et al: Autoimmune haemolytic anaemia associated with ovarian dermoid cyst. *J. Clin. Path.* 21:626–630, 1968.
17. Bakos, L., Storck, R., and Silveira Netto, E.: Congenital macular urticaria pigmentosa complicated by massive nodular mastocytosis with systemic involvement. *Brit. J. Derm.* 87:635–641, 1972.
18. Balsam, A., Bernstein, G., Goldman, J., et al: Ectopic ACTH syndrome associated with carcinoma of the colon. *Gastroenterology* 62:636–641, 1972.
19. Banerjee, R. N. and Narang, R. M.: Haematological changes in malignancy. *Brit. J. Haemat.* 13:829–843, 1967.
20. Bariéty, M. and Coury, C.: L'ostéo-arthropathie hypertrophiante pneumique et les dysacromélies d'origine thoracique. Aspects anatomo-cliniques et évolutifs. A propos de 25 cas. *Semaine des Hôpitaux* 25:1681–1726, 1950.
21. Barnes, R. D.: Thymic neoplasms associated with refractory anemia. *Guy's Hosp. Rep.* 114:73–82, 1965.

22. Barraclough, M. A., Jones, J. J., and Lee, J.: Product of vasopressin by anaplastic oat cell carcinoma of the bronchus. *Clin. Sci.* 31:135–144, 1966.
23. Barry, K. G. and Crosby, W. H.: Auto-immune hemolytic anemia arrested by removal of an ovarian teratoma: a review of the literature and report of a case. *Ann. Intern. Med.* 47:1002–1007, 1957.
24. Bastrup-Madsen, P. and Sondergaard Petersen, H.: Monoclonal gammapathy with the M-component behaving like Donath-Landsteiner haemolysin. *Scand. J. Haemat.* 8:81–85, 1971.
25. Bauer, K. et al: *Wien. Klin. Wschr.* 86:766, 1974. Intravasale Denaturierung von Plasmaproteinen.
26. Bauer, M., Bergström, R., Ritter, B., et al: Macroglobulinemia Waldenström and motor neuron syndrome. *Acta Neurol. Scand.* 55:245–250, 1977.
27. Baumann, G., Lopez-Amor, E., and Dingman, J. F.: Plasma arginine vasopressin in the syndrome of inappropriate antidiuretic hormone secretion. *Amer. J. Med.* 52:19–24, 1972.
28. Baylin, S. B., Gann, D. S., and Hsu, S. H.: Clonal origin of inherited medullary thyroid carcinoma and pheochromocytoma. *Science* 193:321–323, 1976.
29. Bazex, A., Salvador, R., Dupré, A., et al: Dermatose psoriasiforme acromélique d'étiologie cancéreuse. *Bull. Soc. Derm. Syph.* 1966, pp. 130–135.
30. Beck, C. and Burger, H. G.: Evidence for the presence of immunoreactive growth hormone in cancers of the lung and stomach. *Cancer* 30:75–79, 1972.
31. Becker, D. J., Sternberg, M. S., and Kalser, M. H.: Hepatoma associated with hypercalcemia. *J.A.M.A.* 186:1018–1019, 1963.
32. Becker, K. L., Cottrell, J., Moore, C. F., et al: Endocrine studies in a patient with a gonadotropin-secreting bronchogenic carcinoma. *J. Clin. End. Metab.* 28:809–818, 1968.
33. Belcher, R. V. and Smith, S. G.: Study of porphyrins present in hepatoma tissue. *Biochemical Journal* 119:16P, 1970.
34. Belchetz, P. E., Brown, C. L., Makin, H. L. J., et al: ACTH, glucagon and gastrin production by a pancreatic islet cell carcinoma and its treatment. *Clin. End.* 2:307–316, 1973.
35. Benson, Jr., R. C., Riggs, B. L., Pickard, et al: Radioimmunoassay of parathyroid hormone in hypercalcemic patients with malignant disease. *Amer. J. Med.* 56:821–826, 1974.
36. Bergen, S. and Schilling, F. J.: Circulating fibrinolysin in a case of prostatic carcinoma with bony metastases. *Ann. Intern. Med.* 48:389–397, 1958.
37. Berger, L. and Sinkoff, M. W.: Systemic manifestations of hypernephroma. *Amer. J. Med.* 22:791–796, 1957.
38. Bergstrand, C. G. and Czar, B.: Paper electrophoretic study of human fetal serum proteins with demonstration of a new protein fraction. *Scand. J. Clin. Lab. Invest.* 9:277–286, 1957.
39. Bernard, J., Lasneret, J., Chome, J., et al: A cytological and histological study of acute premyelocytic leukaemia. *J. Clin. Path.* 16:319–324, 1963.
40. von Bernheimer, H., Ehringer, H., Heistracher, P., et al: Biologisch aktives, nicht metastasierendes Bronchuscarcinoid mit Linksherzsyndrom. *Wien. Klin. Wschr.* 72:867–873, 1960.
41. Bernier, J. J., Rambaud, J. C., Cattan, D., et al: Diarrhoea associated with medullary carcinoma of the thyroid. *Gut* 10:980–985, 1969.
42. Bigner, D. D., Olson, W. H., and McFarlin, D. E.: Peripheral polyneuropathy, high and low molecular weight IgM, and amyloidosis. *Arch. Neurol.* 24:365–373, 1971.
43. Bill, A. H.: Studies of the mechanism of regression of human neuroblastoma. *J. Pediat. Surg.* 3:727–734, 1968.
44. Bing, J., Fog, M., and Neel, A. V.: Reports of a third case of hyperglobulinemia with affection of the central nervous system on a toxi-infectious basis. *Acta Med. Scand.* 91:409, 1937.
45. Blair, A. J., Hawker, C. D., and Utiger, R. D.: Ectopic hyperparathyroidism in a patient with metastatic hypernephroma. *Metabolism* 22:147–154, 1973.
46. Bleicher, S. J. and Chowdury, F.: Hypoglycemia associated with intrathoracic mesothelioma: studies and a proposed mechanism. *J. Clin. Invest.* 48:9a, 1969.
47. Blix, S. and Aas, K.: Giant haemenangioma, thrombocytopenia, fibrinogenopenia, and fibrinolytic activity. *Acta Med. Scand.* 169:63–70, 1961.

48. Block, J. B.: Lactic acidosis in malignancy and observations on its possible pathogenesis, in Hall, T. (ed.): *Paraneoplastic Syndromes. Annals of the New York Academy of Sciences,* vol. 230, 1974, pp. 94–102.

49. Bloom, S. R.: An enteroglucagon tumour. *Gut* 13:520–523, 1972.

50. Bloom, S. R., Polak, J. M., and Pearse, A. G. E.: Vasoactive intestinal peptide and watery-diarrhoea syndrome. *Lancet* 1973:2, 14–16.

51. Bodel, P.: Tumors and fever, in Hall, T. (ed.): *Paraneoplastic Syndromes. Annals of the New York Academy of Sciences,* vol. 230, 1974, 6–13.

52. Bogaert, R., de Loecker, W., and Tverdy, G.: Amyloidose secondaire au carcinome renal. Etude clinique de trois cas.

53. von Bonsdorff, B., Groth, H., and Packalén, T.: On the presence of high-molecular crystallizable protein in multiple myeloma. *Folia Haemat.* 39:184, 1938.

54. Böttiger, L. E.: Fever of unknown origin. IV. Fever in carcinoma of the kidney. *Acta Med. Scand.* 156:477–485, 1957.

55. Böttiger, L. E., Blanck, C., and von Schreeb, T.: Renal carcinoma: an attempt to correlate symptoms and findings with the histopathologic picture. *Acta Med. Scand.* 180:329–338, 1966.

56. Böttiger, L. E. and Ivemark, B. I.: The structure of renal carcinoma correlated to its clinical behaviour *J. Urol.* 81:512–514, 1959.

57. Boudin, Georges: *Les syndromes paranéoplasiques.* Edition du Concours Médical, Paris, 1962.

58. Brain, L. and Adams, R. D.: A guide to the classification and investigation of neurological disorders associated with neoplasms, in Brain, L. and Norris, F., Jr. (eds.): *The Remote Effects of Cancer on the Nervous System.* Grune & Stratton, New York, 1965.

59. Brain, L. and Norris, F. B., Jr.: The remote effects of cancer on the nervous system. *Contemporary Neurology Symposia,* vol. 1. *Grune & Stratton,* New York, 1965.

60. Brain, L. and Wilkinson, M.: Subacute cerebellar degeneration in patients with carcinoma, in *The Remote Effects of Cancer on the Nervous System.* Grune & Stratton, New York, Brain, L. and Norris, F., Jr. (eds.): 1965, pp. 17–23.

61. Braunstein, G. D. Bridson, W. E., Glass, A., et al: In vivo and in vitro production of human chorionic gonadotropin and alpha-fetoprotein by a virilizing hepatoblastoma. *J. Clin. End.* 35:857–862, 1972.

62. Braunstein, G. D., Vogel, C. L., Vaitukaitis, J. L., et al: Ectopic production of human chorionic gonadotrophin in Ugandan patients with hepatocellular carcinoma. *Cancer* 32:223–226, 1973.

63. Bıaunstein, G. D., Vaitukaitis, J. L., Carbone, P. P., et al: Ectopic production of human chorionic gonadotrophin by neoplasms. *Ann. Intern. Med.* 78:39–45, 1973.

64. Brereton, H. D., Halushka, P. V., et al: Indomethacin responsive hypercalcemia in the patient with renal cell carcinoma. *New Engl. J. Med.* 291:83–85, 1974.

65. Brewer, Jr., H. B., Fairwell, T., Rittel, W., et al: Recent studies on the chemistry of human, bovine and porcine parathyroid hormones. *Amer. J. Med.* 56:759–766, 1974.

66. Brogren, N., Duner, H., Hamrin, B., et al: Urticaria pigmentosa (mastocytosis). A study of nine cases with special reference to the excretion of histamine in urine. *Acta Med. Scand.* 163:223–233, 1959.

67. Browder, A. A., Huff, J. W., and Petersdorf, R. G.: The significance of fever in neoplastic disease. *Ann. Intern. Med.* 55:932–942, 1961.

68. Brown, J. and Winkelmann, R. K.: Acanthosis nigricans: a study of 90 cases. *Medicine* 47:33–51, 1968.

69. DeBruyère, M., Sokal, G., Devoitille, J. M., et al: Autoimmune haemolytic anaemia and ovarian tumour. *Brit. J. Haemat.* 20:83–94, 1971.

70. Buckle, R. M., McMillan, M., and Mallinson, C.: Ectopic secretion of parathyroid hormone by a renal adenocarcinoma in a patient with hypercalcemia. *Brit. Med. J.* 4:724–726, 1970.

71. Burgert, E. O., et al: Intra-abdominal, angiofollicular lymph node hyperplasia (plasma-cell variant) with an antierythropoietic factor. *Mayo Clin. Proc.* 50:542–546, 1975.

72. Burgus, R., Ling, N., Butcher, M., et al: Primary structure of somatostatin, a hypothalamic peptide that inhibits the secretion of pituitary growth hormone. *Proc. Nat. Acad. Sci.* 70:684–688, 1973.

73. Cantrell, E. G.: Nephrotic syndrome cured by removal of gastric carcinoma. *Brit. Med. J.* 2:739–740, 1969.

74. Carey, R. M., Orth, D. N., and Hartmann, W. H.: Malignant melanoma with ectopic production of adrenocorticotropic hormone: Palliative treatment with inhibitors of adrenal steroid biosynthesis. *J. Clin. End.* 36:482–487, 1973.

75. Castaigne, P.: Données actuelles sur la leucoencéphalopathie multifocale progressive. Role oncogène du virus Papova chez l'homme? *Bull. l'Academie National de Médecine* 159:660–664, 1975.

76. Castaldi, P. A. and Penny, R.: A macroglobulin with inhibitory activity against coagulation factor VIII. *Blood* 35:370–376, 1970.

77. Cattan, D., Belaiche, J., Milhaud, G., et al: Cancer bronchique à petites cellules, syndrome de Schwartz-Bartter et hyperthyrocalcitonémie. *La Nouvelle Presse Médicale* 3:2391–2394, 1974.

78. Chadfield, H. W. and Khan, A. U.: Acquired hypertrichosis lanuginosa. *Transactions of the St. John's Hospital Dermatological Society* 56:30–34, 1970.

79. Chamberlain, Z. D.: Carotid body tumour associated with diarrhaea and abdominal pain. *Proc. Roy. Soc. Med.* 54:227–228, 1961.

80. Chey, W. Y. and Brooks, F. P. (eds.): *Endocrinology of the Gut.* Charles B. Slack, Thorofare, N.J., 1974.

81. Chisholm, G. D.: Nephrogenic ridge tumors and their syndromes, in Hall, T. (ed.): *Paraneoplastic Syndromes. Annals of the New York Academy of Sciences,* vol. 230, 1974, pp. 403–423.

82. Chisholm, G. D. and Roy, R. R.: The systematic effects of malignant renal tumours. *Brit. J. Urol.* 43:687–700, 1971.

83. Church, R. E. and Crane, W. A. J.: A cutaneous syndrome associated with islet-cell carcinoma of the pancreas. *Brit. J. Derm.* 79:284–286, 1967.

84. Cohn, D. V., MacGregor, R. R., Chu, L. L., et al: Biosynthesis of proparathyroid hormone and parathyroid hormone. *Amer. J. Med.* 56:767–773, 1974.

85. Conn, J. W., Cohen, E. L., Laucas, C. P., et al: Primary reninism. Hypertension, hyperreninemia, and secondary aldosteronism due to renin-producing juxtaglomerular cell tumors. *Arch. Intern. Med.* 130:682–696, 1972.

86. Coombes, R. C., Hillyard, C. J., Greenberg, P. B., et al: Plasma immunoreactive calcitonin in patients with non-thyroid tumours. *Lancet* 1:1080–1088, 1974. (See also *Cancer* 38:2111–2120, 1976.)

87. Cooper, M. R., Cohen, H. J., Huntley, C. C., et al: A monoclonal IgM with antibodylike specificity for phospholipids in a patient with lymphoma. *Blood* 43:493–504, 1974.

88. Cornbleet, M., Bondy, P. K., and Powles, T. J.: Fatal irreversible hypercalcemia in breast cancer. *Brit. Med. J.* 1:145, 1977.

89. Coscia, M., Brown, R. D., Miller, M., et al: Ectopic production of antidiuretic hormone (ADH), adrenocorticotrophic hormone (ACTH) and beta-melanocyte stimulating hormone (beta-MSH) by an oat-cell carcinoma of the lung. *Amer. J. Med.* 62:303–307, 1977.

90. da Costa, C. R., Dupont, E., Hamers, R., et al: Nephrotic syndrome in bronchogenic carcinoma: report of two cases with immunochemical studies. *Clinical Nephrology* 2:245–251, 1974.

91. Costanza, M. E., Pinn, V., Schwartz, R. S., et al: Carcinoembryonic antigen-antibody complexes in a patient with colonic carcinoma and nephrotic syndrome. *New Engl. J. Med.* 289:520–522, 1973.

92. Costanzi, J. J. and Coltman, C. A., Jr.: Kappa chain cold precipitable immunoglobulin G (IgG) associated with cold urticaria. *Clin. Exp. Immunol.* 2:167, 1967.

93. Cottrell, J. C., Becker, K. L., Matthews, M. J., et al: The histology of gonadotropin-secreting bronchogenic carcinoma. *Amer. J. Clin. Path.* 52:720–725, 1969.

94. Croft, P. B. and Wilkinson, M.: Carcinomatous neuromyopathy: its incidence in patients with carcinoma of the lung and of the breast. *Lancet* 1:184–188, 1963.

95. Croughs, R. J. M., Eastham, W. N., Hackeng, W. H. L., et al: ACTH and calcitonin secreting medullary carcinoma of the thyroid. *Clin. End.* 1:157–171, 1972.

96. Cullen, D. R. and Tomlinson, B. E.: Carcinoma with multiple ectopic hormone secretion and associated myopathy. *Postgraduate Medical Journal* 44:472–491, 1968.

97. Curth, H. O.: Cutaneous manifestations associated with malignant internal disease, in Fitzpatrick, T. B., et al (eds.): *Dermatology in General Medicine.* McGraw-Hill, New York, 1971, p. 1567.

98. Dabek, J. T.: Bronchial carcinoid tumour with acromegaly in two patients. *J. Clin. End. Metab.* 38:329–333, 1974.

99. Dailey, J. E. and Marcuse, P. M.: Gonadotropin secreting giant cell carcinoma of the lung. *Cancer* 24:388–396, 1969.

100. Damon, A., Holub, D. A., Melicow, M. M., et al: Polycythemia and renal carcinoma. *Amer. J. Med.* 25:182–197, 1958.

101. Daniels, A. C., Chokroverty, S., Barron, K. D.: Thalamic degeneration, dementia, and seizures: inappropriate ADH secretion associated with bronchogenic carcinoma. *Arch. Neurol.* 21:15–24, 1969.

102. Davidson, C. S.: Hepatocellular carcinoma and erythrocytosis. *Seminars in Hematology,* Vol. 13. April 1976.

103. Davies-Jones, G. A. B. and Esiri, M. M.: Neuropathy due to amyloid in myelomatosis. *Brit. Med. J.* 2:444, 1971.

104. Davis, L. E. and Drachman, D. B.: Myeloma neuropathy. Successful treatment of two patients and review of cases. *Arch. Neurol.* 27:507–511, 1972.

105. Degos, R., Touraine, R., Belaich, S., et al: Syndrome de Bazex (dermatose psoriasiforme acromélique d'étiologie cancéreuse). *Soc. Derm. Syph.* March 16, 1968, pp. 348–349.

106. Delmonte, L. and Liebelt, R. A.: Granulocytosis-promoting extract of mouse tumour tissue: partial purification. *Science* 148:521–523, 1965.

107. Delmonte, L. and Liebelt, R. A.: Stimulation of granulocyte release and spleen colony forming potential by partially purified extracts of mouse tumor and normal mammalian kidney. abstracted. *Fed. Proc.* 25:232, 1966.

108. Demis, D. J., Walton, M. D., and Higdon, R. S.: Histaminuria in urticaria pigmentosa. *Arch. Derm.* 83:127–38, 1961.

109. Denny-Brown, D. E.: Primary sensory neuropathy with muscular changes associated with carcinoma. *J. Neurol. Neurosurg. Psychiat.* 11:73–87, 1948.

110. van Dijk, E.: Erythema gyratum repens. *Dermatologica* 123:301–310, 1961.

111. van Dijk, E.: Ichthyosiform atrophy of the skin associated with internal malignant diseases. *Dermatologica* 127:413–428, 1963.

112. Dobson, R. L., Young, M. R., and Pinto, J. S.: Palmar keratoses and cancer. *Arch. Dermat.* 92:553–558, 1965.

113. Drusin, L. W., Litwin, S. D. Armstrong, D. and Webster B. P.: Macroglobulinemia in a patient with a chronic biologic false positive test for syphilis. *Amer. J. Med.* 56:429–432, 1974.

113a. Duchon, J. and Pechan, Z.: The biochemical and clinical significance of melanogenuria. *Ann. N.Y. Acad. Sci.* 100:1048–1068, 1963.

114. Duperrat, B., Pringuet, R., and David, V.: Erythema gyratum repens. *Bull. Soc. Fr. Derm. Syph.* 68:578–582, 1961.

115. Dupont, A., et al.: β-Endorphin: Stimulation of growth hormone release in vivo. *Proc. Natl. Acad. Sci.* 74:358–359, 1977.

116. Eaton, L. M. and Lambert, E. H.: Electromyography and electric stimulation of nerves in diseases of motor unit: observations on myasthenic syndrome associated with malignant tumors. *J.A.M.A.* 163:1117–1124, 1957.

117. Editorial: Renal tubular syndromes. Immunologic disorders and cancer. *Ann. Intern. Med.* 67:213–214, 1967.

118. Edmonds, M., Molitch, M., Pierce, J. G., et al: Secretion of alpha subunits of luteinizing hormone (LH) by the anterior pituitary. *J. Clin. End. Metabol.* 41:551–555, 1975.

119. Ehrlich, M., Goldstein, M., and Heinemann, H. O.: Hypocalcemia, hypoparathyroidism and osteoblastic metastases. *Metabolism* 12:516–526, 1963.

120. Elias, E., Polak, J. M., Bloom, S. R., et al: Pancreatic cholera due to production of gastric inhibitory polypeptide. *Lancet* 2:791–793, 1972.

121. Ellis, J. M.: Urticaria pigmentosa: report of case with autopsy. *Arch. Path.* 48:426–435, 1949.

122. Ellison, M. L., Hillyard, C. J. Bloomfield, G. A., et al: Ectopic hormone production by bronchial carcinomas in culture. *Clin. End.* 5 (suppl.):397s–406s, 1976.

123. *Endocrine and Nonendocrine Hormone-producing Tumors*. Year Book Medical Publishers, Chicago, 1973.

124. Engel, A. and von Euler, U. S.: Diagnostic value of increased urinary output of noradrenaline and adrenaline in phaeochromocytoma. *Lancet* 2:387, 1950.

125. Engel, F. L. and Kahana, L.: Cushing's syndrome with malignant corticotropin-producing tumor—Remission and relapse following subtotal adrenalectomy and tumor resection. *Amer. J. Med.* 34:726–734, 1963.

126. Engelman, K., Lowenberg, W. and Sjoerdsma, A.: Inhibition of serotonin synthesis by para-chlorophenylalanine in patients with the carcinoid syndrome. *New Engl. J. Med.* 277:1103–1108, 1967.

127. Engle, R. L., Jr. and Wallis, L. A.: Multiple myeloma and the adult Fanconi syndrome. *Amer. J. Med.* 22:5–23, 1957.

128. Entwistle, C. C., Fentem, P. H., and Jacobs, A.: Red-cell aplasia with carcinoma of the bronchus. *Brit. Med. J.* 2:1504–1506, 1964.

129. von Euler, U. and Pernow, B. (eds.): *Substance P*. Raven Press, New York, 1977.

130. Evaldsson, U., Ertekin, C., Ingvar, D. H., et al: Encephalopathia hypercalcemia. A clinical and electroencephalographic study in myeloma and other disorders. *J. Chron. Dis.* 22:431–449, 1969.

131. Evans, D. J. and Azzopardi, J. G.: Distinctive tumours of bone and soft tissue causing acquired vitamin-D-resistant osteomalacia. *Lancet* 1:353–354, 1972.

132. Fahey, J. L., Barth, W. F., and Solomon, A.: Serum hyperviscosity syndrome. *J.A.M.A.* 192:120–123, 1965.

133. Fahey, R. J.: Unusual lekocyte response in primary carcinoma of the lung. *Cancer* 4:930–935, 1951.

134. Faiman, C., Colwell, J. A., Ryan, R. J., et al: Gonadotropin secretion from a bronchogenic carcinoma. *New Engl. J. Med.* 277:1395–1399, 1967.

135. Falck, B., Ljungberg, O., and Rosengren, R.: On the occurrence of monoamines and related substances in familial medullary thyroid carcinoma with phleochromocytoma. *Acta Path. Microbiol. Scandinav.* 74:1–10, 1968.

136. Farhangi, M. and Osserman, E. F.: Myeloma with xanthoderma due to an IgG monoclonal anti-flavin antibody. *New Engl. J. Med.* 294:177–183, 1976.

137. Feldman, P., Shapiro, L. Pick, A. I., et al: Scleromyxedema. A dramatic response to Melphalan. *Arch. Derm.* 99:41–56, 1969.

138. Feyserter, F.: The clear cell system: The peripheral endocrine (paracrine) glands, in Taylor S. (ed.): *Endocrinology*. Heinemann, London, 1971, p. 134.

139. Fisher, J. W.: Erythropoietin: pharmacology, biogenesis, and control of production. *Pharmacol. Rev.* 24:459–508, 1972.

140. Fisher, M. M., Hochberg, L. A., and Wilensky, N. D.: Recurrent thrombophlebitis in obscure malignant tumor of the lung. *J.A.M.A.* 147:1213–1216, 1951.

141. Fishman, W. H., Inglis, N. R., Green, S., et al: Immunology and biochemistry of Regan isoenzyme of alkaline phosphatase in human cancer. *Nature* 219:697–699, 1968.

142. Fitzpatrick, T. B., et al.: The melanocyte system, in Fitzpatrick, T. B. et al (eds.): *Dermatology in General Medicine*. McGraw-Hill, New York, 1971.

143. Flint, G. L., Flam, M. and Soter, N. A.: Acquired ichthyosis. A sign of non-lymphoproliferative malignant disorders. *Arch. Derm.* 111:1446–1447, 1975.

144. Forssell, J.: Nephrogenous polycythaemia. *Acta. Med. Scand.* 161:169–179, 1958.

145. Franchimont, P., Gaspard, U., Reuter, A., et al: Polymorphism of protein and polypeptide hormones. *Clin. End.* 1:315–336, 1972.

146. Fretzin, D. F.: Malignant down. *Arch. Derm.* 95:294–297, 1967.

147. Fusco, F. D. and Rosen, S. W.: Gonadotropin-producing anaplastic large-cell carcinomas of the lung. *New Engl. J. Med.* 275:508–515, 1966.

148. Gammel, J. A.: Erythema gyratum repens. Skin manifestations in patient with carcinoma of breast. *A.M.A. Arch. Derm. Syph.* 66:494–505, 1952.

149. Ganguly, A., Gribble, J., Tune, B., et al: Renin-secretin Wilms' tumor with severe hypertension. Report of a case and brief review of renin-secreting tumors. *Ann. Intern. Med.* 79:835–837, 1973.

150. Gardner, B., Graham III, W. P., Gordan, C. S., et al: Calcium and phosphate metabolism in patients with disseminated breast cancer: Effect of androgens and of prednisone. *J. Clin. End.* 23:1115–1124, 1963.

151. Geary, C. G., Platts, M. M., and Stewart, A. K.: Hypokalemia of unknown aetiology complicating Hodgkin's disease. *Brit. Med. J.* 2:507–508, 1966.

152. George, J. M., Capen, C. C., and Phillips, A. S.: Biosynthesis of vasopressin in vitro and ultrastructure of a bronchogenic carcinoma. *J. Clin. Invest.* 51:141–148, 1972.

153. Gewirtz, G. and Yalow, R. S.: Ectopic ACTH production in carcinoma of the lung. *J. Clin. Invest.* 53:1022–1032, 1974.

154. Gleeson, M. H., Bloom, S. R., Polak, J. M., et al: Endocrine tumour in kidney affecting small bowel structure, motility, and absorptive function. *Gut* 12:773–782, 1971.

155. Goetzl, E. J. and Austen, K. F.: Structural determinants of the eosinophil chemotactic activity of the acidic tetrapeptides of eosinophilic chemotactic factor of anaphylaxis. *J. Exp. Med.* 1424–1437, 1976.

156. Gold, P. and Friedman, S.: Specific carcinoembryonic antigens of the human digestive system. *J. Exp. Med.* 122:467–481, 1965.

157. Goldberg, D. M. and Ellis, G.: An assessment of serum acid and alkaline phosphatase determinations in prostatic cancer with a clinical validation of an acid phosphatase assay utilizing adenosine 3'-monophosphate as substrate. *J. Clin. Path.* 27:140–147, 1974.

158. Golde, D. W., Schambelan, M., Weintraub, B. D., et al: Gonadotropin-secreting renal carcinoma. *Cancer* 33:1048–1053, 1974.

159. Goldsmith, R. S. and Ingbar, S. H.: Inorganic phosphate treatment of hypercalcemia of diverse etiologies. *N. Engl. J. Med.* 274:1–7, 1966.

160. Goodnight, Jr., S. H.: Bleeding and intravascular clotting in malignancy: a review, in Hall, T. (ed.): *Paraneoplastic Syndromes. The New York Academy of Sciences.* 1974, pp. 271–288.

161. Gordan, G. S. and Roof, B. S.: Humors from tumors: diagnostic potential of peptides. *Ann. Intern. Med.* 76:501–502, 1972.

162. Gordan, G. S., Roof, B. S., and Tomkins, G. M.: Endocrine manifestations of malignant disease. *Calif. Med.* 116:43–51, 1972.

163. Gordon, P. R., Huvos, A. G., and Strong, E. W.: Medullary carcinoma of the thyroid gland. *Cancer* 31:915–924, 1973.

164. Gössner, W. and Korting, G. W.: Metastasierendes Inselzellkarzinoma vom A-Zelltyp bei einem Fall von Pemphigus foliaceus mit Diabetes renalis. *Dtsch. Med. Wschr.* 85:434–437, 1960.

165. Gottschalk, R. G. and Furth, J.: Polycythemia with features of Cushing's syndrome produced by luteomas. *Acta Hematologica* 5:100–123, 1951.

166. Graham, and Helvik, Bowen's disease and its relation to cancer. *Arch. Derm.* 80:137–159, 1959.

167. Graham, D. Y. Johnson, C. D., Bentlif, P. S., et al: Islet cell carcinoma, pancreatic cholera, and vasoactive intestinal peptide. *Ann. Intern. Med.* 83:782–785, 1975.

168. Grahame-Smith, D. G.: *The Carcinoid Syndrome.* Heinemann, London, 1972.

169. Greenberg, P. B., Martin, T. J., and Sutcliff, H. F.: Synthesis and release of parathyroid hormone by a renal carcinoma in cell culture. *Clinical Science* 45:183–191, 1973.

170. Greenberg, P. L. and Creger, W. P.: The anemia of chronic disorders due to renal cell carcinoma: ferrokinetic and morphologic documentation of its surgical correction. *Amer. J. Med. Sci.* 261:265–269, 1971.

171. Greenfield, L. J. and Shelley, W. M.: The spectrum of neurogenic tumours of the sympathetic nervous system: maturation and adrenergic function. *J. Nat. Cancer. Inst.* 35:215–226, 1965.

172. Gregory, R. A. and Tracy, H. J.: Isolation of two "big gastrins" from Zollinger-Ellison tumor tissue. *Lancet* 3:797–799, 1972.

173. Greider, M. H., Rosai, J., and McGuigan, J. E.: The human pancreatic islet cells and their

tumours. II. Ulcerogenic and diarrheogenic tumors. *Cancer* 33:1423–1443, 1974.

174. Gutman, A. B., Lloyd Tyson, T. and Gutman, E. B.: Serum calcium, inorganic phosphorus and phosphatase activity. *Arch. Intern. Med.* 57:379, 1936.

175. Hagen, C. and McNeilly, A. S.: Changes in circulating levels of LH, FSH, LH β- and α-subunit after gonadotropin-releasing hormone, and of TSH, LH β- and α-subunit after thyrotropin-releasing hormone. *J. Clin. End. Metab.* 41:466–470, 1975.

176. Håkansson, et al.: Substance P-like immunoreactivity in intestinal carcinoid tumours, in von Euler, U. and Pernow, B. (eds.): *Substance P.* Raven Press, N.Y. 1977.

176a Hall, T. (ed.): *Paraneoplastic Syndromes. Annals of the New York Academy of Sciences,* vol. 230, 1974.

177. Hamilton, B. P. M., Upton, G. V., and Amatruda, T. T., Jr.: Evidence for the presence of neurophysin in tumors producing the syndrome of inappropriate antidiuresis. *J. Clin. End. Metab.* 35:764–767, 1972.

178. Hamilton, J. R., Radde, I. C., and Johnson, G.: Diarrhea associated with adrenal ganglioneuroma. *Amer. J. Med.* 44:453–463, 1968.

179. Hammar, H.: Erythema annulare centrifugum coincident with Epstein-Barr virus infection in an infant. *Acta Paediatr. Scand.* 63:788–792, 1974.

180. Hammarsten, J. F. and O'Leary, J.: The features and significance of hypertrophic osteoarthropathy. *A.M.A. Arch. Intern. Medicine* 99:431–441, 1957.

181. Hamrin, B.: Release of histamine in urticaria pigmentosa. *Lancet* 1:867–868, 1957.

182. Hamrin, B.: Sustained hypotension and shock due to an adrenaline-secreting phaeochromocytoma. *Lancet* 2:123–124, 1962.

183. Harboe, M., Fölling, I., Hanger, O. A. J. and Bauer, K.: Sudden death caused by interaction between a macroglobulin and a divalent drug. *Lancet* I:285–288, 1976.

184. Harper, P. S., Harper, R. M. J., and Howel-Evans, A. W.: Carcinoma of the oesophagus with tylosis. *Quart. J. Med.* 39:317–326, 1970.

185. Harris, M., Jenkins, M. V., Bennett, A., et al: Prostaglandins, bone resorption and hypercalcemia. *New Engl. J. Med.* 289–865, 1973.

186. Havard, C. W. H.: Thymic tumours and refractory anaemia. *Series Haematologica* 5:18–50, 1965.

187. Havard, C. W. H. and Bodley Scott, R.: Thymic tumour and erythroblastic aplasia. *Brit. J. Hemat.* 6:178–190, 1960.

188. Hawley, P. R., Johnston, A. W., and Rankin, J. T.: Association between digital ischaemia and malignant disease. *Brit. Med. J.* 3:208–212, 1967.

189. Hayduk, K. and Kaufmann, W.: Ektope paraneoplastische Endokrinopathien mit Störungen des Wasser- und Elektrolythaushaltes. *Klin. Wschr.* 51:361–376, 1973.

190. Hegedus, S. I. and Schorr, W. F.: Acquired hypertrichosis lanuginosa and malignancy. *Arch. Derm.* 106:84–88, 1972.

191. Heitz, Ph., Steiner, H., Halter, F., et al: Multihormonal, amyloid-producing tumour of the islets of Langerhans in a twelve-year-old boy. *Virchows Arch. Abt. A. Path. Anat.* 353:312–324, 1971.

192. Hellström, K. E. and Hellström, I.: Immunity to neuroblastomas and melanomas. *Ann. Rev. Med.* 23:19–38, 1972.

193. Herzberg, J. J., ed.: *Cutane paraneoplastische Syndrome.* Verlag, Stuttgart, 1971.

194. Herzberg, J. J., Potjan, K., and Gebauer, D.: Hypertrichosis lanuginosa (et terminalis) aquitsia als paraneoplastisches Syndrom. *Archiv für klinische und experimentelle Dermatologie* 232:176–186, 1968.

195. Hildebrand, J.: Les neuropathies paranéoplasiques. *Revue Médicale de Bruxelles* 28:349–358, 1972.

196. Hildebrand, J. and Coërs, C.: The neuromuscular function in patients with malignant tumours: Electromyographic and histological study. *Brain* 90:67–82, 1967.

197. Hills, E. A.: Adenocarcinoma of the bronchus with Cushing's syndrome, carcinoid syndrome, neuromyopathy and urticaria. *Brit. J. Dis. Chest* 62:88–92, 1968.

198. Hirata, Y., Yamamoto, H., Matsukura, S., et al: In vitro release and biosynthesis of tumor ACTH in ectopic ACTH-producing tumors. *J. Clin. End. Metab.* 41:106–114, 1975.

199. Hirst, E. and Robertson, T. I.: The syndrome of thymoma and erythroblastopenic anemia.

Medicine 46:225–264, 1967.

200. Hochleitner, H., Bartsch, G., and Zelger, J.: Erythema gyratum repens bei Bronchuscarcinom. *Hautarzt* 21:116–119, 1970.
201. Hökfelt, T., Elde, R., Johansson, O., et al: Immunohistochemical evidence for the presence of somatostatin, a powerful inhibitory peptide, in some primary sensory neurons. *Neuroscience Letters* 1:231–235, 1975.
202. Hollis, W. C.: Hypertrophic osteoarthropathy secondary to upper-gastrointestinal-tract neoplasm. *Ann. Intern. Med.* 66:125–130, 1967.
203. Horai, T., Nishihara, H., Tateishi, R., et al: Oat cell carcinoma of the lung simultaneously producing ACTH and serotonin. *J. Clin. End. Metab.* 37:212–219, 1973.
204. Horwitz, A. and McKelway, W. P.: Polycythemia associated with uterine myomas. *J.A.M.A.* 158:1360–1361, 1955.
205. Howel-Evans, W., McConnell, R. B., Clarke, C. A., et al: Carcinoma of the oesophagus with keratosis palmaris et plantaris (tylosis). *Quart. J. Med.* 27:413–429, 1958. (See also *Quart. J. Med.* 39:317–333, 1970.)
206. Huckstep, R. L. and Bodkin, P. E.: Vagotomy in hypertrophic pulmonary osteoarthropathy associated with bronchial carcinoma. *Lancet* August 16, 1958, p. 343–345.
207. Humphreys, S. R., Holley, K. E., Smith, L. H., et al: Mesenteric angiofollicular lymph node hyperplasia (lymphoid hamartoma) with nephrotic syndrome. *Mayo Clin. Proc.* 50:317–321, 1975.
208. Humphreys, G. H., II, and Southworth, H.: Aplastic anemia terminated by removal of mediastinal tumor. *Amer. J.M.Sc.* 210:501–510, 1945.
209. Hung, W., Blizzard, R. M., Migeon, C. J., et al: Precocious puberty in a boy with hepatoma and circulating gonadotropin. *J. Pediat.* 63:895–903, 1963.
210. Imawari, M., Akatsuka, N., Ishibashi, M., et al: Syndrome of plasma cell dyscrasia, polyneuropathy, and endocrine disturbances. *Ann. Intern. Med.* 81:490–493, 1974.
211. Imhof, J. W., Vleugels Schutter, G. J. N., Hart, H. C., et al: Monoclonal gammopathy (IgG) and chronic ulcerative dermatitis (phagedenic pyoderma). *Acta Med. Scand.* 186:289–292, 1969.
212. Isaacson, N. H. and Rapoport, P.: Eosinophilia in malignant tumors; its significance. *Ann. Intern. Med.* 25:893–902, 1946.
213. Isawa, T., Okubo, K., Konno, K., et al: Cushing's syndrome caused by recurrent malignant bronchial carcinoid. *Amer. Rev. Resp. Dis.* 108:1200–1204, 1973.
214. Ishizaka, K. and Dayton, D. (eds.): *Biological Role of the Immunoglobulin E System.* U.S. Department of Health, 1972, pp. 266–268.
215. Isler, P. and Hedinger, C.: Metastasierendes Dünndarmcarcinoid mit schweren vorwiegend das rechte Herz betreffenden Klappenfehlern—ein eigenartiger Symptomenkomplex? *Schweiz. Med. Wschr.* 83:4–7, 1953.
216. Iwashita, H., Argyrakis, A., Lowitzsh, K., et al: Polyneuropathy in Waldenstrom's macroglobulinaemia. *J. Neurol. Sciences* 21:341–354, 1974.
217. Jablonska, S., Stachow, A., and Dabrowska, H.: Rapports entre la pyodermite gangreneuse et le myélome. *Ann. Derm. Syph.* 94:121, 1967.
218. James, K., Fudenberg, H., Epstein, W. L., et al: Studies on a unique diagnostic serum globulin in papular mucinosis (lichen myxedematosus). *Clin. Exp. Immunol.* 2:153, 1967.
219. Jaworski, Z. F. and Wolan, C. T.: Hydronephrosis and polycythemia. *Amer. J. Med.* 34:523–534, 1963.
220. Johnson, L. A., Ancona, V. C., Johnson, T., et al: Primary osteogenic sarcoma of the kidney. *J. Urol.* 104:528–531, 1970
221. Jorpes, E., Holmgren, H., and Wilander, O.: Heparin in vessel walls and in the eyes. *Z. Mikr. Anat. Forsch.* 42:279, 1937.
222. Joseph, R. R., Day, H. J., Sherwin, R. M., et al: Microangiopathic haemolytic anaemia associated with consumption coagulopathy in a patient with disseminated carcinoma. *Scand. J. Haematol.* 4:271–282, 1967.
223. Kakkar, V. V., Howe, C. T., Nicolaides, A. N., et al: Deep vein thrombosis of the leg. *Amer. J.*

Surg. 120:527–530, 1970.

224. Kaplan, L. I., Sokoloff, L., Murray, F., et al: Sympathicoblastoma with metastases associated with the clinical picture of Cushing's syndrome—Report of a case. *Arch. Neurol. Psychiat.* 62:696–698, 1949.

225. Kasabach, H. H. and Merritt, K. K.: Capillary hemangioma with extensive purpura; report of case. *Amer. J. Dis. Child.* 59:1063–1070, 1940.

226. Käser, H.: Die Bedeutung der 3-Methoxy-4-Hydroxy-Mandelsäure (MHMS) für die Differentialdiagnostik neuraler Tumoren im Kindesalter. *Schweiz. Med. Wschr.* 91:586–589, 1961.

227. Käser, H. and von Studnitz, W.: Urine of children with sympathetic tumours. *Amer. J. Dis. Child.* 102:199–204, 1961.

227a. Kellen, J. A. and Malkin, A., Alkali Resistant Hemoglobin in Cancer Patients, AACR abstracts, 363, 1978.

228. Kawamura, J., Garcia, J. H., and Kamijyo, Y.: Cerebellar hemangioblastoma: histogenesis of stroma cells. *Cancer* 31:1528–1540, 1973.

229. Keczkes, K. and Barker, D. J.: Malignant hepatoma associated with acquired hepatic cutaneous porphyria. *Arch. Dermat.* 112:78–82, 1976.

230. Kennedy, B. J., Tibbetts, D. M., Nathanson, T., et al: Hypercalcemia, a complication of hormone therapy of advanced breast cancer. *Cancer Res.* 13:445–459, 1953.

231. Kennedy, J. H., Williams, M. J., and Sommers, S. C.: Cushing's syndrome and cancer of the lung—Pituitary Crooke cell hyperplasia in pulmonary oatcell carcinoma. *Ann. Surg.* 160:90–94, 1964.

232. Keutmann, H. T., Aurbach, G. D., Dawson, B. F., et al: Isolation and characterization of the bovine parathyroid isohormones. *Biochemistry* 10:2779–2787, 1971.

233. Killander, A., Killander, J., Philipson, L., et al: A monoclonal gamma-macroglobulin complexing with lecithin, in Killander, J. (ed.): *Nobel Symposium* 3. *Gamma Globulins* Interscience, New York, 1967, p. 359.

234. Kimball, K. G.: Amyloidosis in association with neoplastic disease. *Ann. Intern. Med.* 55:958–974, 1961.

235. Kither, K., Masopust, J., and Rádl, J.: Fetal α-globulin of bovine serum differing from fetuin. *Biochem. Biophys. Acta* 160:135–137, 1968.

236. Klein, D. C. and Raisz, L. G.: Prostaglandins: Stimulation of bone resorption in tissue culture. *Endocrinology* 86:1436–1440, 1970.

237. Klein, L. A., Rabson, A. S., and Worksman, J.: In vitro synthesis of vasopressin by lung tumour cells. *Surgical Forum* 20:231–233, 1969.

238. Klockars, M., Azar, H. A., Hermida, R., et al: The relationship of lysozyme to the nephropathy in chloroleukemic rats and the effects of lysozyme loading on normal rat kidneys. *Cancer Research* 34:47–60, 1974.

239. Knill-Jones, R. P., Buckle, R. M., Parsons, V., et al: Hypercalcemia and increased parathyroid-hormone activity in a primary hepatoma. *New Engl. J. Med.* 282:704–708, 1970.

240. Koblenzer, P. J. and Baker, L.: Hypertrichosis lanuginosa associated with diazoxide therapy in prepubertal children: a clinicopathologic study. *Ann. N.Y. Acad. Sci.* 150:373–381, 1968.

241. Kock, N. G., Darle, N., and Dotevall, G.: Inhibition of intestinal motility in man by glucagon given intraportally. *Gastroenterology* 53:88–92, 1967.

242. Krain, L. S. and Bierman, S. M.: Pemphigus vulgaris and internal malignancy. *Cancer* 33:1091–1099, 1974.

243. Kreisberg, R. A., Hershman, J. M., Spenney, J. G., et al: Biochemistry of extrapancreatic tumor hypoglycemia. *Diabetes* 19:248–258, 1970.

243a. Krook, G., and Waldenström, J. G., Relapsing Annular Erythema and Myeloma Successfully Treated with Cyclophosphamide. *Acta Med. Scand.* 203:289–292, 1978.

244. Lafferty, F. W.: Pseudohyperparathyroidism. *Medicine* 45:247–260, 1966.

245. Lai A Fat, R. F. M., Suurmond, D., Rádl, J., et al: Scleromyxoedema (lichen myxoedematosus) associated with a paraprotein, IgG_1 of type kappa. *Brit. J. Derm.* 88:107–116, 1973.

246. Lambert, E. H. and Rooke, E. D.: Myasthenic state and lung cancer, in Brain, L. and Norris, F., Jr. (eds): *The Remote Effects of Cancer on the Nervous System.* Grune & Stratton, New York, 1965, pp. 67–80.

247. Landaw, S. A.: Hemolytic anemia as a complication of carcinoma: case report and review of the literature. *J. Mount Sinai Hosp.* 31:167–178, 1964.

248. Law, D. H., Liddle, G. W., Scott, H. W., Jr., et al: Ectopic production of multiple hormones (ACTH, MSH and gastrin) by a single malignant tumor. *New Engl. J. Med.* 273:292–296, 1965.

249. Le Bourhis, J., Fève, J.-R., Besançon, C., et al: Neuropathie périphérique avec infiltration amyloide des nerfs au cours d'une macroglobulinémie de Waldenström. *Revue Neurologique* 111·474–478, 1964.

250. LeDouarin, N. M. and Teillet, M. A.: The migration of neural crest cells to the wall of the digestive tract in avian embryos. *J. Embryol. Exp. Morphol.* 30:31–48, 1973.

251. Lee, J. C., Yamauchi, H., and Hopper Jr., J.: The association of cancer and the nephrotic syndrome. *Ann. Intern. Med.* 64:41–51, 1966.

252. Leichter, S. B., Pagliara, A. S., Greider, M. H., et al: Uncontrolled diabetes mellitus and hyperglucagonemia associated with an islet cell carcinoma. *Amer. J. Med.* 58:285–293, 1975.

253. Lerner, A. B. and Watson, C. J.: Studies of cryoglobulins. I. Unusual purpura associated with the presence of a high concentration of cryoglobulin (cold precipitable serum globulin). *Amer. J. Med. Sci.* 214:410–415, 1947.

254. Levin, J. and Conley, Lockard, C.: Thrombocytosis associated with malignant disease. *Arch. Intern. Med.* 114:497–500, 1964.

255. Liddle, G. W., Givens, J. R., Nicholson, W. E., et al: The ectopic ACTH syndrome. *Cancer Research* 25:1057–1061, 1965.

256. Liddle, G. W., Nicholson, W. E., Island, D. P., et al: Clinical and laboratory studies of ectopic humoral syndromes. *Recent Progr. Hormone Res.* 25:283–305, 1969.

257. Lieberman, J. S., Borrero, J., Urdaneta, E., et al: Thrombophlebitis and cancer. *J.A.M.A.* 177:542–545, 1961.

258. Lightman, S. L. and Bloom, S. R.: Cure of insulin-dependent diabetes mellitus by removal of a glucagonoma. *Brit. Med. J.* 1:367–368, 1974.

259. Lindgärde, F., and Zettervall, O.: Characterization of a calcium-binding IgG myeloma protein. *Scand. J. Immun.* 3, 277–285, 1974.

260. Lipsett, M. B.: Hormonal syndromes associated with neoplasia, in Levine, R. and Luft, R. (eds.): *Advances in Metabolic Disorders.* vol 3. Academic Press, New York, 1968, pp. 125–152.

261. Lipsett, M. B., Odell, W. D., Rosenberg, L. E., et al: Humoral syndromes associated with nonendocrine tumors. *Ann. Intern. Med.* 61:733–756, 1964.

262. Little, J. M.: Potassium imbalance and rectosigmoid neoplasia. *Lancet* 1:302–303, 1964.

263. Ljungberg, O.: Argentaffin cells in human thyroid and parathyroid and their relationship to C-cells and medullary carcinoma. *Acta Path. Microbiol. Scand.* Section A. 80:589–599, 1972.

264. Ljunggren, E., Holm, S., Karth, B., et al: Some aspects of renal tumors with special reference to spontaneous regression. *J. Urol.* 82:553–557, 1959.

265. Lobell, M., Boggs, D. R., and Wintrobe, M. M.: The clinical significance of fever in Hodgkin's disease. *Arch. Intern. Med.* 117:335–342, 1966.

266. Locke, S. A.: *Arch. Intern. Med.* 15:659–713, 1915.

267. Logothetis, J., Silverstein, P., and Coe, J.: Neurologic aspects of Waldenström's macroglobulinemia. *Arch. Neurol.* 3:564–573, 1960.

268. MacDonald, R. A. and Robbins, S. L.: The significance of nonbacterial thrombotic endocarditis: an autopsy and clinical study of 78 cases. *Ann. Intern. Med.* 46:255–273, 1957.

269. Mackenzie, A. H. and Scherbel, A. L.: Connective tissue syndromes associated with carcinoma. *Geriatrics* 18:745–753, 1963.

270. Macris, N. T., Capra, J. D., Frankel, G. J., et al: A lambda light chain cold agglutinin-cryomacroglobulin occurring in Waldenström's macroglobulinemia. *Amer. J. Med.* 48:524–529, 1970.

271. McArthur, J. W., Toll, G. D., Russfield, A. B., et al: Sexual precocity attributable to ectopic gonadotropin secretion by hepatoblastoma. *Amer. J. Med.* 54:390–403, 1973.

272. McFadzean A. J. S., Todd, D., and Tso, S. C.: Erythrocytosis associated with hepatocellular carcinoma. *Blood* 29:808–811, 1967.

273. McFadzean, A. J. S. and Yeung, R. T. T.: Further observations on hypoglycemia in hepatocellular carcinoma. *Amer. J. Med.* 47:220–235, 1969.

274. McGavran, M. H., Unger, R. H., Recant, L., et al: A glucagon-secreting alpha-cell carcinoma of the pancreas. *New Engl. J. Med.* 274:1408–1413, 1966.

275. McGovern, G. P., Miller, D. H., and Robertson, E.: A mental syndrome associated with lung carcinoma. *Arch. Neurol. Psychiat.* 81:341–347, 1959.

276. McLellan, G., Baird, C. W., and Melick, R.: Hypercalcemia in an Australian hospital adult population. *Med. J. Aust.* 2:354–356, 1968.

277. Magnusson, B.: Lichen ruber bullosus and tumours in internal organs. *Dermatologica* 134:166–172, 1967.

278. Mahmoud et al: *Clinical Research* 25:519A, 1977.

279. Maldonado, J. E., Sheps, G. G., Bernatz, P. H., et al: Renal arteriovenous fistula. A reversible cause of hypertension and heart failure. *Amer. J. Med.* 37:499–513, 1964.

280. Mallinson, C. N., Bloom, S. R., Warin, A. P., et al: A glucagonoma syndrome. *Lancet* 2:1–5, 1974.

281. Mandema, E.: Over het multipel myeloom, het solitaire plasmocytoom en de macroglobulinaemie. Groningen 1956.

282. Mannheimer, I. H.: Hypercalcemia of breast cancer. *Cancer* 18:679–691, 1965.

283. Marks, L. J., Steinke, J., Podolsky, S., et al: Hypoglycemia associated with neoplasia. *Paraneoplastic Syndromes. Annals of the New York Academy of Sciences,* 230:147–160, 1974.

284. Marks, V., Auld, W. H. R., and Barr, J. B.: Carcinoma of stomach and other non-pancreatic lesions as causes of spontaneous hypoglycemia. *Brit. J. Surg.* 52:925–928, 1965.

285. Marks, V., Samols, E., and Bolton, R.: Hyperinsulinism and Cushing's syndrome. *Brit. Med. J.* May 29, 1965, p. 1419–1420.

286. Mason, A. M. S., Ratcliffe, J. G., Buckle, R. M., et al: ACTH secretion by bronchial carcinoid tumours. *Clin. End.* 1:3–25, 1972.

287. Mathys, S., Ziegler, W. H., and Francke, C.: Bilaterales Phäochromozytom—medulläres Schilddrüsenkarzinom mit Cushing-Syndrom. *Schweiz. Med. Wschr.* 102:798–803, 1972.

288. Mavligit, G. M., Cohen, J. L., and Sherwood, L. M.: Ectopic production of parathyroid hormone by carcinoma of the breast. *New Engl. J. Med.* 285:154–156, 1971.

289. Mayr, A. C., Dick, H. J., Nagel, G. A., et al: Thrombozytose bei malignen Tumoren. *Schweiz. Med. Wschr.* 103:1626–1629, 1973.

290. Meador, C. K., Liddle, G. W., Island, D. P., et al: Cause of Cushing's syndrome in patients with tumors arising from "nonendocrine" tissue. *J. Clin. End.* 22:693–703, 1962.

291. Melvin, K. E. W.: Esophageal polyp with hypertrophic osteoarthropathy. *Proc. Roy. Soc. Med.* 58:576–577, 1965.

292. Melvin, K. E. W., Tashjian, Jr., A. H., Cassidy, C. E., et al: Cushing's syndrome caused by ACTH- and calcitonin-secreting medullary carcinoma of the thyroid. *Metabolism* 19:831–838, 1970.

293. Merskey, C. A., Johnson, A. J., Pert, J. H. et al: Pathogenesis of fibrinolysis in defibrination syndrome: effect of heparin administration. *Blood* 24:701–715, 1964.

294. Miller, D.: Heparin precipitability of the macroglobulin in a patient with Waldenström's macroglobulinemia. *Amer. J. Med.* 28:951, 1960.

295. Miller, M. and Moses, A. M.: Urinary antidiuretic hormone in polyuric disorders and in inappropriate ADH syndrome. *Ann. Intern. Med.* 77:715–721, 1972.

296. Mirabel, L.: Migrating thrombophlebitis associated with malignant neoplasms. *Canad. Med. Ass. J.* 70:34–38, 1954.

297. Moertel, C. G., Beahrs, O. H., Woolner, L. B., et al: "Malignant carcinoid syndrome" associated with non-carcinoid tumours. *New Engl. J. Med.* 273:244–248, 1965.

297a. Möller, H., Eriksson, S., Holen, O., and Waldenström, J. G., Complete Reversibility of Paraneoplastic Acanthosis Nigricans after Operation, *Acta Med. Scand.* 203:245–246, 1978.

297b. Möller, H., Waldenström, J. G., and Zettervall, O., Pyoderma Gangraenosum (Dermatitis Ulcerosa) and Monoclonal (IgA) Globulin Healed after Melphalan Treatment, *Acta Med. Scand.* 203:293–296, 1978.

298. Moestrup, T. and Hägerstrand, I.: Canalicular activity of phosphatase in human liver biopsy specimens—a paramalignant manifestation?

299. More, I. A. R., Jackson, A. M., MacSween, R. N. M.: Renin-secreting tumor associated with hypertension. *Cancer* 34:2093–2102, 1974.

300. Morgan, A. G., Walker, W. C., Mason, M. K., et al: A new syndrome associated with hepatocellular carcinoma. *Gastroenterology* 63:340–345, 1972.

301. Morley, J. B. and Schwieger, A. C.: The relation between chronic polyneuropathy and osteosclerotic myeloma. *J. Neurol. Neurosurg. Psychiat.* 40:432–442, 1967.

302. Morris, R. C., Jr. and Fudenberg, H. H.: Impaired renal acidification in patients with hypergammaglobulinamia. *Medicine* 46:57–69, 1967.

303. Morton, D. L., Itabashi, H. H., and Grimes, O. F.: Nonmetastatic neurological complications of bronchogenic carcinoma: The carcinomatous neuromyopathies. *J. Thoracic and Cardiovascular Surgery* 51:14–29, 1966.

304. Muggia, F. M., Heinemann, H. O., Farhangi, M., et al: Lysozymuria and renal tubular dysfunction in monocytic and myelomonocytic leukemia. *Amer. J. Med.* 47:351–366, 1969.

305. Müller, W. and Schubothe, H.: Symptomatische hämolytische Anämien bei Dermoidcysten. *Folia Haemat. Neue Folge* 2:321–336, 1958.

306. Mundy, G. R., Raisz, L. G., Cooper, R. A., et al: Evidence for the secretion of an osteoclast stimulating factor in myeloma. *New Engl. J. Med.* 291:1041–1046, 1974.

307. Murphy, G. P., Allen, L. E., Staubitz, R. J., et al: Erythropoietin levels in patients with Wilms' tumor. *N.Y. State J. Med.* February, 1972, pp. 487–489.

308. Murray, J. S., Paton, R. R., and Pope, C. E., II: Pancreatic tumor associated with flushing and diarrhea. *New Engl. J. Med.* 264:436–439, 1961.

309. Murray-Lyon, I. M., Cassar, J., Coulson, R., et al: Further studies on streptozotocin therapy for a multiple-hormone-producing islet cell carcinoma. *Gut* 12:717–720, 1971.

310. Mutter, R. D., Tannenbaum, M., and Ultmann, J. E.: Systemic mast cell diseases. *Ann. Intern. Med.* 59:887–906, 1963.

311. Narayan, O., Penney, Jr., J. B., Johnson, R. T., et al: Etiology of progressive multifocal leukoencephalopathy. *New Engl. J. Med.* 289:1278–1282, 1973.

312. Nathanson, L. and Fishman, W. H.: New observations on the Regan isoenzyme of alkaline phosphatase in cancer patients. *Cancer* 27:1388–1397, 1971.

313. Nichols, J., Warren, J. C., and Mantz, F. A.: ACTH-like excretion from carcinoma of the ovary. *J.A.M.A.* 182:713–718, 1962.

314. Nick, J., Contamin, F., Brion, S., et al: Macroglobulinémie de Waldenström avec neuropathie amyloide. *Revue Neurologique* 109:21–30, 1963.

315. Nilsson, I.-M., Sjoerdsma, A., and Waldenström, J. G.: Antifibrinolytic activity and metabolism of E-aminocaproic acid in man. *Lancet* 1:1322–1326, 1960.

316. Norman, T. and Otnes, B.: Intestinal ganglioneuromatosis, diarrhoea and medullary thyroid carcinoma. *Scand. J. Gastroent.* 4:553–559, 1969.

317. Nyman, M., Skölling, R., and Steiner, H.: Acquired macrocytic anemia and hemoglobinopathy—A paraneoplastic manifestation? *Amer. J. Med.* 48:792–797, 1970.

318. Olefsky, J., Kempson, R., Jones, H., et al: "Tertiary" hyperparathyroidism and apparent "cure" of vitamin-D-resistant rickets after removal of an ossifying mesenchymal tumor of the pharynx. *New Engl. J. Med.* 286:740–745, 1972.

318a. Old, J., Langley, J., Wood, W. G., Clegg, J. B., and Weatherall, D. J. Molecular Basis for

Acquired Hemoglobin H. Disease. Nature 269:524–25, 1977.

319. Omenn, G. S.: Ectopic hormone syndromes associated with tumors in childhood. *Pediatrics* 47:613–622, 1971.

320. O'Neal, L. W., Kipnis, D. M., Luse, S. A., et al: Secretion of various endocrine substances by ACTH-secreting tumors—gastrin, melanotropin, norepinephrine, serotonin, parathormone, vasopressin, glucagon. *Cancer* 21:1219–1232, 1968.

321. Opsahl, R.: Thymuskarcinom og aplastisk anemia. *Nord. Med.* 2:1835–1936, 1939.

322. Orth, D. N.: Establishment of human malignant melanoma clonal cell lines that secrete ectopic ACTH. *Nature (New Biol)* 242:26–28, 1973.

323. Orth, D. N., Nicholson, W. E., Mitchell, W. M., et al: Biologic and immunologic characterization and physical separation of ACTH and ACTH fragments in the ectopic ACTH syndrome. *J. Clin. Invest.* 52:1756–1769, 1973.

324. Osserman, E. F.: Lysozymuria in renal and non-renal diseases, in: Manuel, Y., Revillard, J. P., and Betuel, H. (eds.): *Proteins in Normal and Pathological Urine.* Basel, S. Karger, 1970, pp. 260–270.

325. Osserman, E. F. and Lawlor, D. P.: Serum and urinary lysozyme (Muramidase) in monocytic and monomyelocytic leukemia. *J. Exp. Med.* 124, 921–952, 1966.

326 -327. Parry, E. H. O.: Arthropathy due to gastric carcinoma. *Brit. Med. J.* 2:1022, 1958.

328. Payne, R. A. and Ryan, R. J.: Human placental lactogen in the male subject. *J. Urol.* 107:99–103, 1972.

329. Pearse, A. G. E. and Polak, J. M.: Neural crest origin of the endocrine polypeptide (APUD) cells of the gastrointestinal tract and pancreas. *Gut* 12:783–788, 1971.

330. Pedersen, K. O.: *Ultracentrifugal Studies on Serum and Serum Fractions.* dissertation. Almquist and Wiksell, Upsala, 1945.

331. Penman, H. G. and Thomson, K. J.: Amyloidosis and renal adenocarcinoma: A post-mortem study. *J. Path.* 107:45–47, 1972.

332. Penney, Jr., J. B., Weiner, L. P., Herndon, R. M., et al: Virions from progressive multifocal leukoencephalopathy: Rapid serological identification by electron microscopy. *Science* 178:60–62, 1972.

333. Perlo, V., Schwab, R. S., and Castleman, B.: Myasthenia gravis and thymoma, in Brain, L. and Norris, F., Jr. (eds.): *The Remote Effects of Cancer on the Nervous System.* Grune & Stratton, New York, 1965, pp. 55–66.

334. Pernow, B. and Waldenström, J. G.: Paroxysmal flushing and other symptoms caused by 5-hydroxytryptamine and histamine in patients with malignant tumours. *Lancet,* 1954, p. 951.

335. Pernow, B. and Waldenström, J. G.: Determination of 5HT, 5HIAA and histamine in 33 cases of carcinoid tumour. *Amer. J. Med.* 23:16–25, 1957.

336. Peters, M. N., Hall, R. J., Cooley, D. A., et al: The clinical syndrome of atrial myxoma. *J.A.M.A.* 230:695–701, 1974.

337. Pevny, I.: Erythema gyratum repens. *Ztschr. Haut-Geschlkr.* 40:260–270, 1966.

338. Plager, J. and Stutzman, L.: Acute nephrotic syndrome as a manifestation of active Hodgkin's disease. *Amer. J. Med.* 50:56–66, 1971.

339. Plimpton, C. H. and Gellhorn, A.: Hypercalcemia in malignant disease without evidence of bone destruction. *Amer. J. Med.* 21:750–759, 1956.

340. Potter, M., Lieberman, R., Hood, L., et al: Structure and serological studies of six IgA myeloma proteins from Balb/c mice that bind phosphoryl choline. *Fed. Proc.* 29, 437, 1970. (See also Chapter in Azar, H. A. and Potter, M. (eds.): *Multiple Myeloma and Related Disorders,* vol. 1. Harper & Row, New York, 1973.)

341. Powell, D., Singer, F. R., Murray, et al: Nonparathyroid humoral hypercalcemia in patients with neoplastic diseases. *New Engl. J. Med.* 289:176–181, 1973.

342. Prader, A., Illig, R., Uehlinger, E., et al: Rachitis infolge Knochentumors. *Helvetica Paediatrica Acta* 5/6:554–565, 1959.

343. Pruzanski, W., Leers, W.-D., and Wardlaw, A. C.: Bacteriolytic and bactericidal activity in monocytic and myelomonocytic leukemia with hyperlysozymemia. *Cancer Research* 33:867–873, 1973.

344. Pruzanski, W. and Platt, M. E.: Serum and urinary proteins, lysozyme (muramidase), and renal dysfunction in mono- and myelomonocytic leukemia. *J. Clin. Invest.* 49:1694–1708, 1970.

345. Pusterla, E. and Hedinger, C.: Die Häufigkeit medullärer Schilddrüsenkarzinome bei ein- und doppelseitigen Phäochromozytomen. *Schweiz. Med. Wschr.* 3:83–87, 1975.

346. Rabson, A. S., Rosen, S. W., Tashjian, et al: Production of human chorionic gonadotrophin in vitro by a cell line derived from a carcinoma of the lung. *J. Natl. Cancer Inst.* 50:669–674, 1973.

347. Rausing, A. and Axelsson, U.: Progressive multifocal leukoencephalopathy in chronic lymphatic leukemia—caused by polyoma virus? *Scand. J. Haemat.* 7:184–194, 1970.

348. Rawlins, M. D., Luff, R. H., and Cranston, W. I.: Pyrexia in renal carcinoma. *Lancet* 1:1371–1373, 1970.

349. Rees, L. H., Bloomfield, G. A., Rees, G. M., et al: Multiple hormones in a bronchial tumor. *J. Clin. End.* 38:1090–1097, 1974.

350. Rees, L. H. and Ratcliffe, J. G.: Ectopic hormone production by non-endocrine tumours. *Clin. End.* 3:263–299, 1974.

351. Revol, MM. L., Viala, J.-J., Revillard, J.-P., and Manuel, Y.: Protéinurie associée à des manifestations paranéoplasiques au cours d'un cancer bronchique. *Lyon Médical* 212:907–916, 1964.

352. Richardson, E. P., Jr.: Progressive multifocal leukoencephalopathy. *New Engl. J. Med.* 265:815–823, 1961.

353. Richardson, E. P., Jr.: Our evolving understanding of progressive multifocal leukoencephalopathy, in Hall, T. (ed.): *Paraneoplastic Syndromes. The New York Academy of Sciences,* vol. 230. 1974, pp. 358–364.

354. Riesen, W., Noseda, G., and Butler, R.: Anti-lipoprotein activity of human monoclonal immunoglobulins. *Vox Sang. (Basel)* 22:420, 1972.

354a. Riesen, W., Rudikoff, S., Oriol, R., et al: An IgM Waldenström with specificity against phosphorylcholine. *Biochemistry* 14:1052–1056, 1975.

355. Robboy, S. J., Colman, R. W., and Minna, J. D.: Pathology of disseminated intravascular coagulation (DIC), analysis of 26 cases. *Human Pathol.* 3:327–343, 1972.

356. Robertson, J. C. and Eeles, G. H.: Syndrome associated with pancreatic acinar cell carcinoma. *Brit. Med. J.* 2:708–709, 1970.

357. Robertson, P. W., Klidjian, A., Harding, L. K., et al: Hypertension due to renin-secreting renal tumour. *Amer. J. Med.* 43:963–976, 1967.

358. Robinson, W. A.: Granulocytosis in neoplasia, in Hall, T. (ed.): *Paraneoplastic Syndromes. Annals of the New York Academy of Sciences,* vol. 230. 1974, 212–218.

359. Robinson, W. A. and Mangalik, A.: The regulation of granulopoiesis: positive feed-back. *Lancet* 2:742–743, 1972.

360. Röckl, H., Knedel, M., and Schröpl, F.: Über das Vorkommen von Paraproteinämie bei Pyodermia ulcerosa serpiginosa (Pyoderma gangraenosum-Dermatitis ulcerosa). *Der Hautarzt* 15:165–171, 1964.

361. Roof, B. S., Carpenter, B., Fink, D. J., et al: Some thoughts on the nature of ectopic parathyroid hormones. *Amer. J. Med.* 50:686–691, 1971.

362. Root, A. W., Bongiovanni, A. M., and Eberlein, W. R.: A testicular-interstitial-cell-stimulating gonadotrophin in a child with hepatoblastoma and sexual precocity. *J. Clin. End. Metab.* 28:1317–1322, 1968.

363. Rorsman, H., Rosengren, A.-M., and Rosengren, E.: A sensitive method for determination of 5-S-cysteinyldopa. *Acta Dermatovener (Stockholm)* 53:248–250, 1973.

364. Rosen, P. and Armstrong, D.: Nonbacterial thrombotic endocarditis in patients with malignant neoplastic diseases. *Amer. J. Med,* 54:23–29, 1973.

365. Rosen, S. W. and Weintraub, B. D.: Ectopic production of the isolated alpha subunit of the glycoprotein hormones. *New Engl. J. Med.* 290:1441–1447, 1974.

366. Rosenstein, B. J. and Engelman, K.: Diarrhea in a child with a catecholamine-secreting ganglioneuroma. *J. Pediat.* 63:217–226, 1963.

367. Rosenthal, R. L.: Acute promyelocytic leukemia associated with hypofibrinogenemia. *Blood* 21:495–508, 1963.

368. Rothman, S.: Studies on melanuria. *J. Lab. Clin. Med.* 27:687, 1942.

369. Rowland, L. P. and Schneck, S. A.: Neuromuscular disorders associated with malignant neoplastic disease. *J. Chron. Dis.* 16:777–795, 1963.

370. Roy, A. D. and Ellis, H.: Potassium secreting tumor of the large intestine. *Lancet* 1:759–760, 1959.

371. Ruoslahti, E., Seppälä, M., Räsänen, J. A., et al: Alpha-fetoprotein and hepatitis B antigen in acute hepatitis and primary cancer of the liver. *Scand. J. Gastroent.* 8:197–202, 1973.

372. Sachs, B. A., Becker, N., Bloomberg, A. E., et al: "Cure" of ectopic ACTH syndrome secondary to adenocarcinoma of the lung. *J. Clin. End.* 30:590–597, 1970.

373. Sackner, M. A., Spivack, A. P., and Balian, L. J.: Hypocalcemia in the presence of osteoblastic metastases. *New Engl. J. Med.* 262:173–176, 1960.

374 -376. Salama, F., Luke, R. G., and Hellebusch, A. A.: Carcinoma of the kidney producing multiple hormones. *J. Urol.* 106:820–722, 1971.

377. Salassa, R. M., Jowsey, J., and Arnaud, C. D.: Hypophosphatemic osteomalacia associated with "nonendocrine" tumors. *New Engl. J. Med.* 283:65–70. 1970.

378. Salmon, S. E.: "Paraneoplastic" syndromes associated with monoclonal lymphocyte and plasma cell proliferation, in Hall, T. (ed.): *Paraneoplastic Syndromes. Annals of the New York Academy of Sciences,* vol. 230, 1974, pp. 228–239.

379. Sandler, M., Scheuer, P. J., and Watt, P. J.: 5-Hydroxytryptophan-secreting bronchial carcinoid tumour. *Lancet* 2:1067–1069, 1961.

380. Schambelan, M., Howes, E. L., Jr., Stockigt, J. R., et al: Role of renin and aldosterone in hypertension due to a renin-secreting tumor. *Amer. J. Med.* 55:86–92, 1973.

381. Schersten, T., Wahlqvist, L., and Johansson, L. G.: Lysosomal enzyme activity in liver tissue from patients with renal carcinoma. *Cancer* 23:608–613, 1969.

382. Schonfeld, A., Babbott, D., and Gunderson, K.: Hypoglycemia and polycythemia associated with primary hepatoma. *New Engl. J. Med.* 265:231–233, 1961.

383. Schteingart, D. E., Conn, J. W., Orth, D. N., et al: Secretion of ACTH and β-MSH by an adrenal medullary paraganglioma. *J. Clin. End.* 34:676–683, 1972.

384. Schwab, P. J. and Fahey, J. L.: Treatment of Waldenström's macroglobulinemia by plasmapheresis. *New Engl. J. Med.* 263:574–579, 1960.

385. Schwartz, W. B., Bennett, W., Curelop, S., et al: A syndrome of renal sodium loss and hyponatremia probably resulting from inappropriate secretion of antidiuretic hormone. *Amer. J. Med.* 23:529–542, 1957.

386. Scott, J. A.: 3,4-dihydroxyphenylalanine (DOPA) excretion in patients with malignant melanoma. *Lancet* 2:861–862, 1962.

387. Scott, R. B., Elmore, S. McD., Brackett, N. C., Jr., et al: Neuropathic joint disease (Charcot Joints) in Waldenström's macroglobulinemia with amyloidosis. *Amer. J. Med.* 54:535–538, 1973.

388. Seyberth, H. W., Segre, G. V., Morgan, J. L., et al: Prostaglandins as mediators of hypercalcemia associated with certain types of cancer. *New Engl. J. Med.* 293:1278–1283, 1975.

389. Shamblin, J. R., Jr., Huff, J. F., Waugh, J. M., et al: Villous adenocarcinoma of the colon with pronounced electrolyte disturbance. *Ann. Surg.* 156:318–326, 1962.

390. Shapiro, M., Nicholson, W. E., Orth, D. N., et al: Differences between ectopic MSH and pituitary MSH. *J. Clin. End.* 33:377–381, 1971.

391. Shaw, G: Bence Jones proteinuria and hyperglobulinemia in a case of renal carcinoma. *Glasgow Med. J.* 35:79–85, 1954.

392. Shelley, W. B. and Hurley, H. J.: An unusual autoimmune syndrome. *Arch. Derm.* 81:889–897, 1960.

393. Shim, W. K. T.: Hemangiomas of infancy complicated by thrombocytopenia. *Amer. J. Surg.* 116:896–906, 1968.

394-396. Silva, O. L., Becker, K. L., Primack, A., et al: Ectopic secretion of calcitonin by oat-cell carcinoma. *New Engl. J. Med.* 290:1122–1124, 1974.

397. Simon, R. and Greene, R. C.: Perirenal hemangiopericytoma. A case associated with hypoglycemia. *J.A.M.A.* 189:155–156, 1964.

398. Sirota, J. H. and Hamerman, D.: Renal function studies in an adult subject with the Fanconi syndrome. *Amer. J. Med.* 16:138–152, 1954.

399. Sjöberg, H. E. and Hjern, B.: Acute treatment with calcitonin in primary hyperparathyroidism and severe hypercalcemia of other origin. *Acta Chir. Scand.* 141:90–95, 1975.

400. Sjoerdsma, A., Weissbach, H., and Udenfriend, S.: A clinical, physiologic and biochemical study of patients with malignant carcinoid (argentaffinoma). *Amer. J. Med.* 20:520–532, 1956.

401. Skarin, A. T., Matsuo, Y., and Moloney, W. C.: Muramidase in myeloproliferative disorders terminating in acute leukemia. *Cancer* 29:1336–1342, 1972.

402. van der Sluis, D. I.: Pyoderma (ecthyma) gangraenosum bij twee patienten met een β 2A-paraproteine in het bloed. *Nederlands Tijdschrift voor Geneeskunde, Jahrgang 109,* No 31:1427–1433, 1965.

403. Sluys Veer, J., Chonfoer, J. C., Querido, A., et al: Metastasising islet cell tumour of pancreas associated with hypoglycaemia and carcinoid syndrome. *Lancet* 1:1416, 1964.

404. Smith, J. K.: The significance of the "protein error" of indicators in the diagnosis of Bence-Jones proteinuria. *Acta Haemat.* 30:144–152, 1963.

405. Solomon, A.: Neurological manifestations of macroglobulinemia, in Brain, L. and Norris, F., Jr. (eds.): *The Remote Effects of Cancer on the Nervous System.* Grune & Stratton, New York, 1965, pp. 112–124.

406. Solomon, A.: Medical progress. Bence-Jones proteins and light chains of immunoglobulins. *New Engl. J. Med.* 294:17–23, 91–98, 1976.

407. Solomon, A. and Fahey, J. L.: Plasmapheresis therapy in macroglobulinemia. *Ann. Intern. Med.* 58:789–800, 1963.

408. Soto, P. J., Jr., Rader, E. S., Martin, J. M., et al: Osteogenic sarcoma of the kidney: report of a case. *J. Urol.* 94:532–535, 1965.

409. deSousa, R. C. and Jenny, M.: Hyponatrémie par dilution dans un cas de carcinome pancréatique. *Schweiz. Med. Wschr.* 94:930–938, 1964.

410. Sparagana, M., Phillips, G., Hoffman, C., et al: Ectopic growth hormone syndrome associated with lung cancer. *Metabolism* 20:730–736, 1971.

411. Stanley, E. R. and Metcalf, D.: Partial purification and some properties of the factors in normal and leukaemic urine stimulating bone marrow colony growth in vitro. *Aust. J. Ex. Biol. Med. Sci.* 46:335–349, 1969.

412. Stauffer, M. H.: Nephrogenic hepatosplenomegaly. *Gastroenterology* 40:694, 1961.

413. Steiner, H., Dahlbäck, O., and Waldenström, J. G.: Ectopic growth-hormone production and osteoarthropathy in carcinoma of the bronchus. *Lancet,* April 13, 1968, pp. 783–785.

414. Stenström, G. and Heedman, P-A.: Clinical findings in patients with hypercalcaemia. *Acta Med. Scand.* 195:473–477, 1974.

415. Stickler, G. B., Hallenbeck, G. A., Flock, E. V., et al: Catecholamines and diarrhea in ganglioneuroblastoma. *Amer. J. Dis. Child.* 104:598–604, 1962.

416. Stillman, A. and Zamcneck, N.: Recent advances in immunologic diagnosis of digestive tract cancer. *Digestive Diseases* 15:1003–1018, 1970.

417. Stolbach, L. L., Krant, M. J., and Fishman, W. H.: Ectopic production of an alkaline phosphatase isoenzyme in patients with cancer. *New Engl. J. Med.* 281:757–762, 1969.

418. Stone, S. P. and Schroeter, A. L.: Bullous pemphigoid and associated malignant neoplasms. *Arch. Dermat.* 111:991–994, 1975.

419. Straub, P. W.: Chronic intravascular coagulation. *Acta Med. Scand.,* suppl. 526, 1971.

420. Straub, P. W.: A case against heparin therapy of intravascular coagulation. *Thrombosis et Diathesis Haemorrhagica* 33:107–112, 1975.

421-422. Straub, P. W., Riedler, G., and Frick, P. G.: Hypofibrinogenaemia in metastatic carcinoma of the prostate: suppression of systemic fibrinolysis by heparin. *J. Clin. Path.* 20:152–157, 1967.

423. von Studnitz, W.: Methodische und klinische Untersuchungen über die Ausscheidung der 3-Methoxy-4-Hydroxymandelsäure im Urin. *Scand. J. Clin. Lab. Invest.,* suppl. 48, 1960.

424. von Studnitz, W.: Ueber die Ausscheidung der 3-Methoxy-4-hydroxyphenylessigsäure (Homovanillinsäure) beim Neuroblastom und anderen neuralen Tumoren. *Klin. Wschr.* 40:163–167, 1962.

425. von Studnitz, W., Käser, H., and Sjoerdsma, A.: Spectrum of catechol amine biochemistry in patients with neuroblastoma. *New Engl. J. Med.* 269:232–235, 1963.

426. von Studnitz, W. and Waldenström, J. G.: Pharmakologische und therapeutische (p-Chlorphenylalanin) Studien über das Carcinoidsyndrom bei 11 Fällen. *Klin. Wschr.* 49:748–755, 1971.

427. Summerly, R.: The figurate erythemas and neoplasia. *Brit. J. Derm.* 76:370–373, 1964.

428. Sundström, C., Lundberg, D., and Werner, I.: A case of thymoma in association with megakaryocytopenia. *Acta Path. Microbiol. Scand.* 80A:487–490, 1972.

429. Sutherland, D. A. and Clark, H.: Hemangioma associated with thrombocytopenia. *Amer. J. Med.* 33:150–157, 1962.

430. Tagnon, H. J., Schulman, P., Whitmore, W. F., et al: Prostatic fibrinolysin. Study of a case illustrating role in hemorrhagic diathesis of cancer of the prostate. *Amer. J. Med.* 15:875–884, 1953.

431. Tashjian, A. H., Levine, L., and Munson, P. L.: Immunochemical identification of parathyroid hormone in non-parathyroid neoplasms associated with hypercalcemia. *J. Exp. Med.* 119:467–484, 1964.

432. Thivolet, J. and Perrot, H.: Pigmentation du visage et des bras, revelatrice d'un dysembriome malin du testicule. In *Cutane paraneoplastische Syndrome* Verlag, Stuttgart, 1971, pp. 43–44.

433. Thompson, M. E., Bromberg, P. A., and Amenta, J. S.: Acid mucopolysaccharide determination. *Amer. J. Clin. Path.* 52:335–339, 1969.

434. Thompson, R. P. H., Nicholson, D. C., Farnan, T., et al: Cutaneous porphyria due to a malignant primary hepatoma. *Gastroenterology* 59:779–783, 1970.

435. Thomson, J. and Stankler, L.: Erythema gyratum repens. Reports of two further cases associated with carcinoma. *Brit. J. Derm.* 82:406–411, 1970.

436. Thorling, E. B.: Paraneoplastic erythrocytosis and inappropriate erythropoietin production. *Scand. J. Hematol.*, suppl. 17, 1972. pp. 1–166.

437. Thorson, Å. H.: Studies on carcinoid disease. *Acta Med. Scand.*, Suppl. 334, 1958.

438. Thorson, Å. H.: Hemodynamic changes during "flush" in carcinoidosis. *Amer. Heart J.* 52:444–461, 1956.

439. Thorson, Å., Biörck, G., Björkman, G., and Waldenström, J. G.: Malignant carcinoid of the small intestine with metastases to the liver, valvular right-sided heart disease, peripheral vasomotor symptoms, bronchoconstriction and an unusual type of cyanosis. *Amer. Heart J.* 47:795–817, 1954.

440. Thorson, Å. and Nordenfelt, O.: Development of valvular lesions in metastatic carcinoid disease. *Brit. Heart J.* 21:243–248, 1959.

441. Tio (Tiong Hoo), Leijnse, B., Jarrett, A., et al: Acquired porphyria from a liver tumour. *Clin. Sci.* 16:517–527, 1957.

442. Tisher, C. G.: Correction of an ADH syndrome by resection of a bronchogenic carcinoma with demonstration of tumor antidiuretic activity. *Clin. Res.* 14:185, 1966.

443. Todesco, S., Terribile, V., Borsatti, A., et al: Primary aldosteronism due to a malignant ovarian tumor. *J. Clin. End. Metab.* 41:809–819, 1975.

444. Trell, E., Rausing, A., Ripa, J., et al: Carcinoid heart disease. *Amer. J. Med.* 54:433–444, 1973.

445. Trivedi, S. A.: Neurilemmoma of the diaphragm causing severe hypertrophic pulmonary osteoarthropathy. *Brit. J. Tuberc. Dis. Chest.* 52:214–217, 1958.

446. Turkington, R. W.: Ectopic production of prolactin. *New Engl. J. Med.* 285:1455–1458, 1971.

447. Tyler, H. R.: Paraneoplastic syndromes of nerve, muscle, and neuromuscular junction, in Hall, T. (ed.): *Paraneoplastic Syndromes. Annals of the New York Academy of Sciences,* vol. 230. 1974, pp. 348–357.

448. Udenfriend, S., Titus, E., and Weissbach, H.: The identification of 5-hydroxy-3-indole acetic acid in normal urine and a method for its assay. *J. Biol. Chem.* 216:499–505, 1955.

449. Upton, G. V. and Amatruda, Jr., T. T.: Evidence for the presence of tumor peptides with

corticotropin-releasing-factor-like activity in the ectopic ACTH syndrome. *New Engl. J. Med.* 285:419–424, 1971.

450. Vertel, R. M., Morse, B. S., and Prince, J. E.: Remission of erythrocytosis after drainage of a solitary renal cyst. *Arch. Intern. Med.* 120:54–58, 1967.

451. Victor, M., Banker, B. Q., and Adams, R. D.: The neuropathy of multiple myeloma. *J. Neurol. Neurosurg. Psychiat.* 21:73–88, 1958.

451a. Videback, A., Mansa, B. and Kjems, E.: Human IgA myeloma protein with antistreptococcal hyaluronidase activity. *Scand. J. Haemat.* 10:181–185, 1973.

452. Voelkel, E. F., Tashjian, A. H., Jr., Franklin, R., et al: Hypercalcemia and tumor-prostaglandins: The VX_2 carcinoma model in the rabbit. *Metabolism* 24:973–986, 1975.

453. Vogel, C. L., Dhru, D., Rorsman, H., et al: Dopa and 5-S-cysteinyldopa in malignant melanoma in Ugandan Africans. *Acta Dermatovener (Stockholm)* 54:19–20, 1974.

454. Voorhess, M. L.: Urinary excretion of DOPA and metabolites by patients with melanoma. *Cancer* 96:146–149.

455. Waddington, R. T.: A case of primary liver tumour associated with porphyria. *Brit. J. Surg.* 59:653–654, 1972.

456. Waldenström, J. G.: Incipient myelomatosis or "essential" fibrinogenopenia—a new syndrome? *Acta Med. Scand.* 117:216–247, 1944.

457. Waldenström, J. G.: Recherches cliniques et physicochimiques sur les hyperglobulinémies. *La Presse Médicale* 57:213–215, 1949.

458. Waldenström, J. G.: Abnormal proteins in myeloma. In Snapper and Dock (ed.): *Advances in Internal Medicine,* vol. 5., 1952, p. 398.

459. Waldenström, J. G.: The production of substances with specific activity ("hormones") by some tumour cells. *Acta Endocrinologica* 17:432–441, 1954.

460. Waldenström, J. G.: Clinical aspects of carcinoid tumours, in *Modern Trends in Gastro-Enterology.* Butterworth, London, 1958, pp. 92–100.

461. Waldenström, J. G.: Die Makroglobulinamie. *Ergeb. Inn. Med. Kinderh.* 9:586–621, 1958.

462. Waldenström, J. G.: Studies on conditions associated with disturbed gamma globulin formation (gammopathies), in *Harvey Lectures.* Academic Press, New York, 1961, pp. 211–231.

463. Waldenström, J. G. Macroglobulinemia, in Luft and Levine (eds.): *Advances in Metabolic Disorders.* 1965, 2:115–158.

464. Waldenström, J. G.: Monoclonal and polyclonal hypergammaglobulinemia: Clinical and biological significance. *The 1965 Abraham Flexner Lectures.* Vanderbilt University Press, Nashville, 1968.

465. Waldenström, J. G.: Eiweissbildung und Tumorzelle. *Behringwerke Mitt. Heft* 49:17–27, 1969.

466. Waldenström, J. G.: *Diagnosis and Treatment of Multiple Myeloma.* Grune & Stratton, New York, 1970.

467. Waldenström, J. G.: Dysproteinaemic neuropathies, in Vinken, P. J. and Bruyn, G. W. (eds.): *Handbook of Clinical Neurology,* vol. 8. North-Holland Publishing, Amsterdam, 1970, pp. 72–76.

468. Waldenström, J. G.: Maladies of derepression. Pathological, often monoclonal, derepression of protein forming templates. *Schweiz. Med. Wschr.* 100:2197–2206, 1970.

469. Waldenström, J. G.: Cutaneous lesions in the dysproteinemias, in Fitzpatrick, T. B., et al (eds.): *Dermatology in General Medicine.* McGraw-Hill, New York, 1971, pp. 1124–1130, 1971.

470. Waldenström, J. G.: Benign monoclonal gammapathies in Azar, H. A. and Potter, M. (eds.): *Multiple Myeloma and Related Disorders,* vol. 1. Harper & Row, New York, 1973, pp. 247–286.

471. Waldenström, J. G.: Specific activities of immunoglobulins produced in monoclonal gammopathy—maladies of derepression. *Europ. J. Cancer* 12:413–418, 1976.

472. Waldenström, J. G.: Carcinoid tumours, in Bockus: *Gastroenterology,* 1976, pp. 473–480.

473. Waldenström, J. G.: Carcinoid tumörer och vasomotorik (tills E. Ljungberg). *Sv. Läkart.* 50:690, 1953.

474. Waldenström, J. G. and Ljungberg, E.: Studies on the functional circulatory influence from metastasizing carcinoid (argentaffine, enterochromaffine) tumours and their possible relation to enteramine production. I. Symptoms of carcinoidosis. *Acta Med. Scand.* 152:293–309, 1955.

475. Waldenström, J. G. and Pernow, B.: Paroxysmal flushing and other symptoms caused by 5-HT

and histamine in patients with malignant tumours. *Lancet* 2:951, 1954.

476. Waldenström, J. G., Pernow, B., and Silwer, H.: Case of metastasizing carcinoma (argentaffinoma?) of unknown origin showing peculiar red flushing and increased amounts of histamine and 5-hydroxy-tryptamine in blood and urine. *Acta Med. Scand.* 156:73–83, 1956.
477. Waldenström, J. G., Winblad, S., Hällén, et al: The occurrence of serological "antibody" reagins or similar γ-globulins in conditions with monoclonal hyperglobulinemia. *Acta Med. Scand.* 176, 619–631, 1964.
478. Waldmann, T. A., Levin, E. H., and Baldwin, M.: The association of polycythemia with a cerebellar hemangioblastoma. The production of an erythropoiesis stimulating factor by the tumor. *Amer. J. Med.* 31:318–324, 1961.
479. Waldmann, T. A., Rosse, W. F., and Swarm, R. L.: The erythropoiesis-stimulating factors produced by tumors. *Ann. N.Y. Acad. Sci.* 149:509–515, 1968.
480. Walker, R. L.: Bence-Jones proteinuria in a case of bronchogenic carcinoma. *Med. J. Australia* 1:248–249, 1962.
481. Walter, R., Rudinger, J., and Schwartz, I. L.: Chemistry and structure-activity relations of the antidiuretic hormones. *Amer. J. Med.* 42:653–677, 1967.
481a.Walton, D. N. and Adams R. D.: *Polymyositis.* Livingstone Ltd., Edinburgh & London, 1958.
482. Warin, A. P.: Necrolytic migratory erythema with carcinoma of pancreas. *Proc. Roy. Soc. Med.* 67:24–26, 1974.
483. Wasserman, S. I., Goetzl, E. J., Ellman, L., et al: Tumor-associated eosinophilotactic factor. *New Engl. J. Med.* 290:420–424, 1974.
484. Wegelins, O. and Pasternack, A.: *Amyloidosis.* Academic Press, London, 1977.
485. Weichert, R. F.: The neural ectodermal origin of the peptide-secreting endocrine glands. *Amer. J. Med.* 49:232–241, 1970.
486. Weinstein, B., Irreverre, F., and Watkin, D. M.: Lung carcinoma, hypouricemia and aminoaciduria. *Amer. J. Med.* 39:520–526, 1965.
487. Weintraub, B. D. and Rosen, S. W.: Ectopic production of human chorionic somatomammatropin by nontrophoblastic cancers. *J. Clin. End. Metab.* 32:94–101, 1971.
488. Wellborn, J. K., Brennan, M. J., and Hathaway, J. C., Jr.: Acute fatal fibrinolysis with gastric carcinoma. *Amer. J. Surg.* 108:344–348, 1964.
489. Westerhausen, M., Kickhöfen, B., Werner, P. et al: Transferrin-Autoimmunsyndrom. Ein neues pathogenetisches Prinzip. *Verhandl. Deutsch. Gesellschaft für innere Medizin.* Verlag, Munich, 1972, p. 901.
490. Westerhausen, M. and Meuret, G.: Transferrin-immune complex disease. *Acta Haemat.* 57:96–101, 1977.
491. Wheat, Jr., M. W. and Ackerman, L. V.: Villous adenomas of the large intestine. *Ann. Surg.* 147:476–487, 1958.
492. Wilkinson, D. S.: Necrolytic migratory erythema with carcinoma of the pancreas. *Transactions of the St John's Hospital Dermatological Society* 59:244–250, 1973.
493. Williams, E. D., Karim, S. M. M., and Sandler, M.: Prostaglandin secretion by medullary carcinoma of the thyroid: a possible cause of the associated diarrhea. *Lancet* 1:22–23, 1968.
494. Williams, E. D., Morales, A. M., and Horn, R. C.: Thyroid carcinoma and Cushing's syndrome. *J. Clin. Path.* 21:129–135, 1968.
495. Williams, E. D. and Sandler, M.: The classification of carcinoid tumours. *Lancet* 1:238–239, 1963.
496. Williams, Jr., R. C.: Dermatomyositis and malignancy: a review of the literature. *Ann. Intern. Med.* 50:1174–1181, 1959.
497. Wills, M. R. and Gowan: Value of plasma chloride concentration and acid-base status in the differential diagnosis of hyperparathyroidism from other causes of hypercalcemia. *J. Clin. Path.* 24:219–227, 1971.
498. Wilson, J. R., Merrick, H., and Vogel, S. D.: Woodward ER: Hyperparthyroid-like state in rabbits with the VX_2 carcinoma: Further studies. *Amer. Surg.* 31:145–152, 1965.

499. Wintrobe, M. M. and Buell, M. V.: Hyperproteinemia associated with multiple myeloma. *Bull. Johns Hopkins Hosp.* 52:156, 1933.

500. Wollheim, F. A.: Clinical and immunochemical observations in cryoglobulinemia with special reference to IgM heterogeneity, in Grubb, R. and Samuelsson, G. (eds.): *Human Anti-Human Gammaglobulins.* Pergamon Press, New York, 1971.

501. Womack, W. S. and Castellano, C. J.: Migratory thrombophlebitis associated with ovarian carcinoma. *Amer. J. Obst. Gynec.* 63:467–469, 1952.

502. Woolling, K. R. and Shick, R. M.: Thrombophlebitis: a possible clue to cryptic malignant lesions. *Proc. Mayo Clin.* 31:227–233, 1956.

503. Worms, R., Pequignot, H., and Israel, L.: Dermatose exfoliante généralisée au cours d'un réticulo-sarcome ganglionnaire de l'âisselle. Disparition rapide et complète des lésions cutanées après excérèse de la tumeur. *Société médicale des hopitaux. Bull. et mem. (Paris)* 67:343–345, 1951.

504. Wrigley, P. F. M., Malpas, J. S., Turnbull, A. L., et al: Secondary polycythaemia due to a uterine fibromyoma producing erythropoietin. *Brit. J. Haemat.* 21:551–555, 1971.

505. Wuhrmann, F.: Über das Coma paraproteinaemicum bei Myelomen und Makroglobulinämien. *Schweiz. Med. Wschr.* 86:623–625, 1956.

506. Wurtman, R. J.: The pineal gland: endocrine interrelationships, in Stollerman, G. H. (ed.): *Advances in Internal Medicine,* vol. 16. Yearbook Medical Publishers, Chicago, 1970, 155–169.

507. Yalow, R. S. and Berson, S. A.: Characteristics of "Big ACTH" in human plasma and pituitary extracts. *J. Clin. End. Metab.* 36:415–423, 1973.

508. Yoshikawa, S., Kawabata, M., Hatsuyama, Y., et al: A typical Vitamin-D resistant osteomalacia. *J. Bone Jt. Surg.* 46-A:998–1007, 1964.

509. Zettervall, O.: Antibody activity in monoclonal immunoglobulin G. *Acta Med. Scand.,* Suppl. 492, 1968.

510. Zilva, J. F. and Nicholson, J. P.: Plasma phosphate and potassium levels in the hypercalcemia of malignant disease. *J. Clin. End. Metab.* 36:1019–1026, 1973.

Index